Animal Agriculture in Developing Countries

Animal Agriculture in Developing Countries

Kurra Venkata Gopaiah

RANDOM PUBLICATIONS
NEW DELHI (INDIA)

Animal Agriculture in Developing Countries

ISBN 978-93-5111-784-1

Published in 2016 in India by

Reprint 2019

RANDOM PUBLICATIONS
4376-A/4B, Gali Murari Lal, Ansari Road
New Delhi-110 002
Phone : +9111-43580356, 011-23289044, 011-43142548
e-mail: sales@randompublications.com,
info@randompublications.com, randomexports@gmail.com

Type Setting by : Friends Media, Delhi-110089
Printed at : Mehra Printers, Delhi-110 092

Preface

The sustainable development of animal agriculture, across the many different ecoregions or agroecological zones that are found in the developing countries, poses many fundamental challenges; challenges to the primary users of livestock, to their extension, research and support service agents (private- and state-sponsored), to local and regional development authorities, to investment banks, government policy makers and their institutional organs and, in the final analysis, to the consumer or user of animal output.

Terms like "Factory Farming" are often used to describe modern animal agriculture. Unfortunately, this terminology seems to de-humanize the practice of raising livestock and poultry, which is not at all true. Farm sizes have gotten bigger for all livestock species and poultry. This is simply a result of efficiencies of size and the public demand for low-cost meat, milk and eggs in the supermarket. We often hear about the loss of the family farm, which in many cases, simply is not true; that family farm has just gotten a lot bigger. This is the same whether we are talking about livestock production or crop production.

The global livestock sector is rapidly changing in response to globalization and growing demand for animal-source foods, driven by population growth and increasing wealth in much of the developing world. The rapid rate of urbanization seen in many countries is not only linked to growing affluence but also gives rise to changes in people's food preferences; usually tending towards greater convenience and higher standards of safety.

Livestock contribute 40 % of the global value of agricultural output and support the livelihoods and food security of almost a 1.3 billion people. The livestock sector is one of the fastest growing parts of the agricultural economy.

The growth and transformation of the sector offer opportunities for agricultural development, poverty reduction and food security gains, but the rapid pace of change risks marginalizing smallholders, and systemic risks to the natural resources and human health must be addressed to ensure sustainability. The book will suit to the needs of students, teachers, scholars and general readers.

– Author

Contents

1

Strategies for Sustainable Animal Agriculture in Developing Countries

STRATEGIES FOR SUSTAINABLE DEVELOPMENT OF ANIMAL AGRICULTURE

The sustainable development of animal agriculture, across the many different ecoregions or agroecological zones that are found in the developing countries, poses many fundamental challenges; challenges to the primary users of livestock, to their extension, research and support service agents (private- and state-sponsored), to local and regional development authorities, to investment banks, government policy makers and their institutional organs and, in the final analysis, to the consumer or user of animal output. These challenges have been confronted by mankind with varying degrees of intensity since man first captured and domesticated wild animals for use. However, it is only in the past 200 years that what may be described as an industrial approach to animal agriculture became manifest. Over the past 40 years, the United Nations (UNDP and FAO), investment agencies such as the World Bank, and many bilateral and non-governmental organizations, all in consort with national governments, have launched development programmes to promote or effect the advancement of livestock production throughout the developing world. There have been many successes, particularly in pig and poultry production (e.g.in Thailand) but, unfortunately, many failures also.

Extensive reviews of livestock investment projects conducted by the World Bank (1985) and the Asian Development Bank (ADB, 1991) paint a gloomy picture of the success of development investment in livestock, particularly in the smallholder sector which farms the vast majority of all livestock in developing countries. This lack of success/impact, in turn, has led to a significant reduction in the amount of investment funds directed to livestock production. Added to this, there is increasing interest in all issues related to the sustainability of agricultural development programmes, including livestock production, not only those factors that may directly effect the environment (e.g.

Co_2, methane), but also a more thorough scrutiny of development inputs and technologies in terms of longterm practical application and impact. The FAO definition of sustainability, embracing both dimensions of long-term development impact, provides the basis on which development strategies. Clearly there is a need to thoroughly analyze the determinants of sustainability in the context of animal agriculture so as to identify and evaluate effective development strategies.

DETERMINANTS OF SUSTAINABILITY

Effective development planning entails a comprehensive understanding of man's need to advance development, a thorough analysis of the technical, economic and social implications of proposed interventions and finally, an assessment of potential environmental impact (long-and short-term) that may influence local, regional or global conditions. These very generalized determinants of sustainability, set a framework within which the development planning process must proceed. It is not the purpose of this chapter to discuss livestock development planning at this broad level other than to highlight that complex +interactions and conflicts of interest can, and very often do, exist among the different components in the development process. The development audiences, i.e., the individual farmer, the extended family or local community and the regional and/or national governments, often have very different perceptions of what development means. Technical interventions may be viable economically, but may not be self-supporting (due to dependence on imports and availability of foreign exchange) and may not be socially acceptable if, for example, they result in increased or conflicting work routines within the farm family. Finally, whereas few farmers in developing countries are acutely concious of the global environmental impact of their actions, they certainly are aware of their local habitats and, generally speaking, respect traditional practices and taboos that condition their farming patterns.

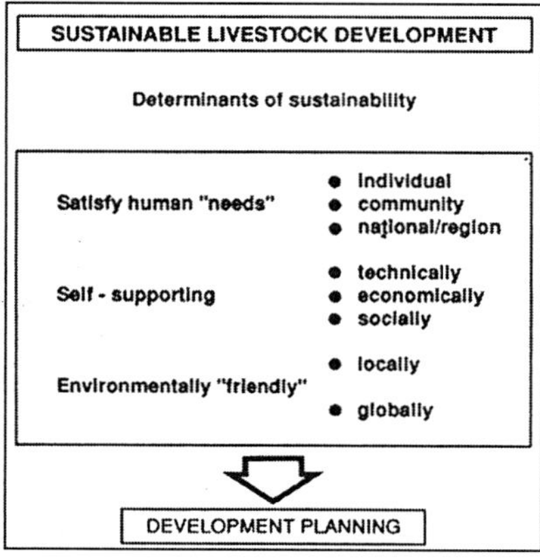

Fig. Determinants of Sustainability

Given the paramount importance of the broad determinants of sustainability and working on the assumption that these are duly considered in framing the development planning process, technical strategies for the advancement of animal agriculture. These will be considered in reference to the classically-defined agricultural development resources, viz., animal and land resources, labour, capital and human enterprise.

THE ANIMAL: PRODUCTION OBJECTIVES AND GENETIC ADVANCE

The development of intensive livestock production in industrial countries over the past 30 years has become synonymous with single purpose breed specialization; consequently, Holstein Friesians for milk production and continental beef breeds such as the Charolais, automatically spring to the minds of livestock development planners throughout industrialized Europe, America and Australasia. Unfortunately, some of these people often export their 'vertical thinking' into livestock development planning in many developing countries; countries to which specialized, single-purpose breeds are not suited and can rarely adapt to the harsh local environments. Market demand, labour and capital costs, together with land availability, have dictated specialized single-purpose animal production systems in the industrial countries; production efficiency is ultimately measured in terms of shelf price and continuity of a standardized, high quality, animal food product.

However, interesting contrasts are found within industrial countries, if, for example, sheep management systems in a land-rich, low population country such as New Zealand, are contrasted with a high population European country such as Britain. In New Zealand, sheep breeding goals are focused on relatively low producing (lambing percentage, 90%-100%), easy-care sheep, where one man can manage approximately 4,000 ewes. In contrast, British lambing percentages in the smaller flocks (200-400 ewes) are targeted at levels approaching 200%.

However, the contrasts in animal breeding objectives between industrial and developing countries are much more stark. Most developing countries exploit the multipurpose use of livestock to produce meat, milk, fibre, draught energy and manure. Several livestock development projects have failed to attract committed farmer participation simply because they were targeted on single-purpose milk or meat output and ignored the farmers primary interest in livestock as providers of animal traction or manure (Lensch, 1985). Further, the bioenergetic efficiency of multipurpose livestock production is often overlooked by livestock planners and, consequently, misplaced development objectives are often pursued. In the Ganges delta of Western Bengal, for example, Odend'hal (1980) studied the energy inputs and output of 3,770 cattle on 2,600 smallholder farms which practised a typical Indian village livestock farming system. Total energy output from the livestock system (Kcal/year), clearly shows that animal manure was a far more important output than milk or meat (calf weight).

This unique example is quoted, not to diminish the general importance of meat and milk in most developing countries, but simply to highlight the need to closely examine how and why farmers use their livestock before deciding on breeding or livestock development objectives in any given situation.

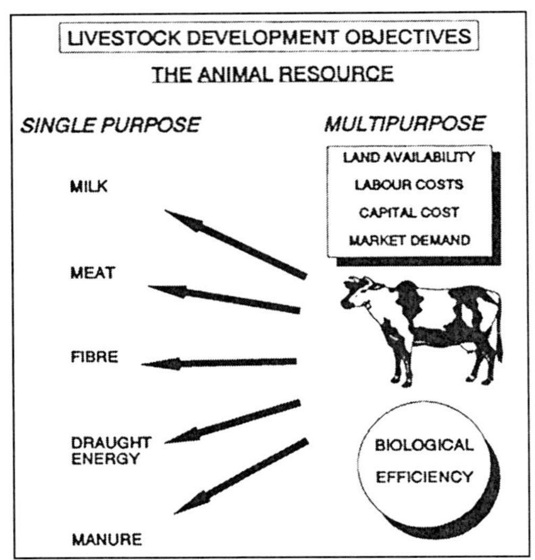

Fig. Livestock Development Objectives

Table. Total energy input and outputs in a traditional village cattle system in West Bengal

p Manure p Traction p Milk p Calves	Kcal/year p3 .93 $\times 10^9$p0.56 $\times 10^9$p0.18 $\times 10^9$p0.01 $\times 10^9$
Total output p Total input p Efficiency	4.68 $\times 10^9$p19.83 $\times 10^9$p23.6%

GENETIC IMPROVEMENT

Perusal of the World Bank's analysis of livestock development projects over the past 25 years clearly shows that many projects focused their attention on the genetic improvement or, in some instances, the replacement of the indigenous animal resource.

It is now widely accepted that the vast majority of such projects failed, and that projects aimed to promote exotic breed propagation failed miserably. Retrospective analysis of these projects would suggest that the fundamental tenets of effective genetic improvement were not adequately considered when these breed improvement projects were designed.

The fundamental conditions that determine the success or failure of genetic improvement programmes are discussed here under four headings:

- Relevance of genetic improvement;
- Breeding objectives;
- Breeding strategies and plans;
- Genetic improvement capacity.

Relevance of Genetic Improvement

The concept of genetic improvement is attractive. However, reality suggests that improvement of the inherent genetic capacity of any animal population beyond the scope of the nutritional- or diseaselimited environment in which the population lives can be meaningless and is often counter-productive. Reports on several development and investment projects, sponsored by UNDP and the World Bank among others, offer tangible proof of this reality. Retrospective analyses of these projects raise the question of when genetic improvement is relevant and potentially useful. McDowell (1989) and Timon and Baber (1989) argue that it makes little sense to initiate genetic improvement programmes in livestock populations where average annual feed intake is less than 1.5 times the animals maintenance feed requirements; ideally it can be argued that feed intake levels should be twice maintenance. The feed intake levels in question refer to the population targeted for improvement, rather than just a nucleus population on a government breeding farm.

The relevance of genetic improvement programmes must also be judged in reference to the adequacy of the breed improvement infrastructure, not only to service and support genetic evaluation and selection within government-supported farms, but also in reference to the effectiveness of regional and national infrastructures through which the propagation and dissemination of the improved germplasm can reach the farmer. This is often very weak or non-existent in developing countries; the failure to develop efficient artificial insemination (AI) services throughout most of the developing world is just one example of weak infrastructure. However, the most important criteria by which the relevance of breed improvement programmes is determined is farmer acceptability; do the changes brought about by a programme adequately benefit the farmers, in terms of perceived market benefit, investment risk, support service availability and general family welfare.

Breeding Objectives

Most animal breeders and, in particular, breeders of ruminant animals, identify breed improvement with increased levels of market-oriented productivity, be that meat or milk. This rationale has emerged from the more advanced countries where input costs (land, labour, capital or support service costs) in animal production are high. Unfortunately, these same breeding objectives, based on the premise that 'big is beautiful', have been exported to many developing country breeding programmes. World Bank (1985) and ADB (1991) project reports provide many examples of failed attempts to achieve 'instant genetic improvement' by importing single-purpose high-producing breeds into unsuited environments.

In the tropics, for example, selection objectives must identify animals that perform well under heat stress and can cope with the seasonally available feed supply and variation in feed quality. Animals must not be evaluated or selected under better managed high-feeding conditions, which are sometimes developed on government farms or research stations, if these conditions are not typical of the prospective farming systems for which the animals are being bred. The classic experiment carried out by Falconer (1960), albeit with experimental animals, highlights this point. His experiments showed that selection for growth rate on an *ad-libitum*plane of nutrition increased appetite, whereas selection for growth rate on a restricted diet increased efficiency of feed utilization.

When genetic improvement within a harsh or limiting environment such as in the tropics of West Africa is considered, high output cannot be considered as the sole criterion of the animals' merit. Brody (1945), comparing a range of animal breeds, and indeed animal species, in terms of productivity per unit of metabolic weight, suggested that there were very small differences between big and small animals in net production efficiency. Taylor et al (1985) largely confirmed Brody's thesis in a set of well-controlled experiments with different breeds of cattle. Comparable experimental evidence from small ruminants is not available but the conclusion is unlikely to be different. In other words, big is not necessarily beautiful in the context of net production efficiency.

It is now well established that specialized single-purpose breeding objectives should not be applied to livestock breeding in the difficult environments of developing countries. Indeed, it may well be questioned in the advanced countries at present as food mountains grow, as market buoyancy and prices are under pressure and as product to energy, feed and labour cost ratios continue to decline.

The basic question at issue is the biological efficiency of multipurpose breeding to produce meat, milk, fibre (skins) and, where relevant, draught power, as compared with the specialized production of one product. Clearly, in the absence of factual data, it is not possible to generalize on this question but it is important to focus attention on the need for a much more comprehensive evaluation of production objectives when livestock breeding strategies are being determined, and particularly when considering difficult environments where economic and social values are very different from those in the more industrialized countries.

Breeding Strategies and Plans

The fundamental basis of a genetic improvement programme, properly focused, depends on selection intensity, selection accuracy and 'population reach'; population reach embraces the relevance of selection objectives and the efficacy of the propagation infrastructure within which the improved germplasm is disseminated to the target farmers.

Crossbreeding: As stated earlier, many genetic improvement programmes have, in the past, been based on the importation of exotic germplasm followed by the initiation of crossbreeding programmes in one form or another; the objective being to harness additive genetic variance in upgrading the indigenous population or the exploitation of heterosis in a continous crossbreeding programme. Choice of exotic breed is certainly important, but extensive reviews by Vaccaro (1990) in South America and McDowell (1983) in Asia raise serious questions about the adaptability of all exotic breeds introduced from temperate countries to tropical conditions.

Certainly, the proportion of 'exotic genes' introduced into indigenous populations is often debated, but the overall benefits of crossbreeding must be seriously assessed and measured, not just on the basis of one trait performance (e.g., lactation yield), but in terms of overall lifetime herd/flock productivity. Measured in these terms, there are few examples of successful and sustained crossbreeding/upgrading programmes in developing countries. The exploitation of heterosis in continuous crossbreeding as opposed to upgrading programmes, is often advocated as a way to rapidly advance animal genetic potential in developing countries. Certainly, there is good evidence of large heterotic effects on some components of animal productivity when exotic temperate and indigenous tropical breeds of cattle are crossbred. Despite this, and for very understandable reasons, sustained crossbreeding programmes are not evident anywhere in the developing world.

The explanation for this has as much or more to do with land use and economics than with genetics. Continuous crossbreeding systems involve three genotypic populations, viz., Breed A and Breed B interbred to produce the crossbred population A × B. In situations where natural mating prevails, this usually requires a sustained economically viable interaction of three sets of farming systems/breeders viz., one set each who breed populations A and B, and a third group who interbreed and farm the A × B crossbred. The sustained integration of three sets of breeders/farmers demands that their respective breeding practices fit to their land resources and compensate them accordingly. Where efficient AI services are available, the basic issue remains the same, except that a sustainable crossbreeding programme with AI demands the complementarity of only two land-use systems instead of three. It is perhaps worth emphasizing that sustainable crossbreeding systems are rare in industrial countries and, indeed, only exist where they 'biologically fit' into traditional land use patterns, e.g., a lowland-upland-mountain land-use pattern as exists in Britain. Molecular genetics (cloning coupled with embryo transfer) may offer new opportunities to exploit crossbreeding in the distant future, but in the immediate term (5 to 10 years) it can be argued that attention should be focused on the genetic improvement of indigenous breeds; in other words, selection within the indigenous breed population, if this is feasible.

Indigenous Breed Improvement: The basis on which to determine the genetic merit of an animal and to make accurate selection decisions have long since been established. Equally, the relative accuracies of different forms of selection, be they pedigree selection, performance testing, progeny or other forms of indexed family or combined selection, are well established. The absence of well-documented genetic parameter estimates makes it difficult in most situations to develop reliable selection indices for livestock breeding in developing countries. However, this is not the only point to be considered. In the majority of countries, the infrastructure and support services necessary to collect the records from which accurate genetic rankings can be made, simply do not exist.

Allied to this is the even greater problem of the very high level of phenotypic variation usually observed in these populations. Coefficients of variation for phenotype can range from 30–60%. This unusually big variation exacerbates the problem of estimating the genetic merit of an individual or group of individuals. An ILCA study (Peters, 1985) has shown that very large numbers of animals are required when comparing goat breeds in the tropics. For example, a group size of 700-1,000 animals is required to detect a difference of 5% (one-tailed test, $\hat{a} = 80\%$) if, as is often the case, the coefficient of variation lies between 35 to 40%. In these medium- to poor-nutritional environments, conventional genetic ranking procedures do not work effectively because of this large variation. McDowell (1983), analyzing cattle records from a number of such environments, concluded that progeny testing, the standard orthodox method of sire evaluation, simply does not work in poor environments. What can be done in these situations? Ironically the lack of a breed infrastructure and this very large variation may be the key to future progress - progress through population screening.

Population Screening: In most breed improvement programmes, within-breed selection is limited to a rather small section of the population - the pedigree or stud breeders. This limits selection pressure, the most important determinant of genetic advance. In most livestock breeds in developing countries a pedigree breed structure does not exist and therefore selection can be considered at the level of the national herd. This may offer an opportunity to change a population genetically that has hitherto been ignored, since;

- Most livestock populations in developing countries have not been subjected to controlled man-directed selection; and

- The large variation observed in these populations may in part be genetic, caused by major genes segregating at low frequencies.

Finally, it should not be forgotten that the major improvement in animal breeding, particularly affecting growth, body composition and conformation, made by livestock breeders in the last century, was achieved using very simple selection procedures. Breeders such as Robert Bakewell and the Colling brothers made dramatic genetic progress by 'screening' local animal populations for better animals as they subjectively assessed them. As a result, they developed the modern European breeds of livestock known today. They exploited the big variation in the then unselected livestock populations with which they worked. Bakewell is quoted as saying "Breed the best to the best and hope for the best". Since Bakewell's time, the genetics of animal production have become more fully understood and current problems can be approached with much greater scientific resolve and understanding.

To explain and hence attempt to capitalize on the unusually large variation that exists in many of the livestock populations in the developing countries, two hypotheses may be considered:

- That genetic variation in these unselected populations is larger than normal to the extent that the total variation is unusually large.
- That major genes with significant effects on production traits are segregating in these populations at low gene frequency.

Evidence to support the first hypothesis partially exists, in that reported estimates of the genetic parameters in livestock populations in developing countries are very similar, if not higher, to those reported for animal populations in the more advanced countries. There is no hard evidence to support the second hypothesis, since it has not been tested, but there is increasing evidence that the screening of national populations may identify major genes. In recent years, at least five examples of major genes have been identified in sheep breeds (Australia, United Kingdom, Ireland, Iceland and Indonesia) and there are currently indications of a major gene controlling milk protein in a French goat breed (Timon, 1990). In any event, the screening of national populations for 'exceptional animals' makes good sense. At worst, it is a means of exercising maximum selection pressure within a population and, of course, it allows the possibility of identifying animals that are carriers of important major genes.

In the difficult environmental conditions which predominate in many developing countries, the question of how to screen large populations in the absence of national recording schemes arises. The only solution to this is to follow the same approach as the pioneer breeders of the last century. The initial screening should be done subjectively; in other words, the breeders are asked to identify 'exceptional animals', if any, in their herds or flocks. These 'exceptional animals' should then be recorded on the farm and compared with a random sample of contemporaries. If their performance approaches or exceeds twice the average level of the herd or flock, then those animals should be acquired, purchased or leased, for more controlled evaluation and breeding on a central test farm or research station. In this way, an open nucleus herd or flock can be established to act as a source of improver stock for the genetic improvement of the target population.

Screening Target: It is deliberately suggested that the selection target is pitched very high, at or near twice the flock average. Assuming a normal distribution for the trait in question and that animals are identified at the very upper end of the distribution, namely three standard deviations (σ) above the mean (μ), it is easy to calculate the resulting selection differentials in terms of the variance or coefficient of variation (cv). If a selection differential ($i = \mu_2 - \mu_1$) is obtained that is equal to or greater than three standard deviations above the mean ($i \geq 3\sigma = 3\mu_1 \times cv$) then it can be calculated that 'exceptional animals' may be defined as being 100 to 200% above average, since the cv ranges from 30 to 70%.

If, on the other hand, a major gene or block of linked genes is segregating in the herd at low frequencies individual animals outside this range may be found. A simulation study of the possible effects of the Booroola gene in the Australian Merino and a specially selected (national screening) sheep flock in Ireland, has shown that coefficients of variation and repeatabilities in these flocks can be very high (60–70%) as a result of the expression of a major gene (Piper and Hanrahan, personal communication).

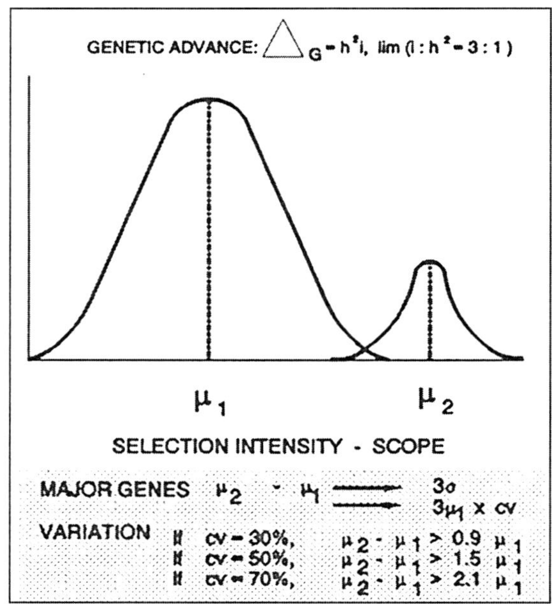

Fig. Selection Intensity Scope

Population screening - Genetic rationale

Examples of successful screening have been reported in Ireland and Britain. In Ireland, the national sheep flock was screened for ewes displaying exceptional litter size, i.e., 3, 4 or more lambs/ewe. The open nucleus flock which was then established had an average litter size of 2.3 lambs/ewe as compared to 1.3 in the national flock. Analysis of litter size distribution and repeatability estimates in this flock suggests the presence of a major gene or block of genes controlling ovulation rate. The Javanese Thin Tailed sheep (JTT) is also showing clear evidence of the existence of a major gene controlling ovulation rate. This latter example is cited as evidence that major genes may and have been found in livestock populations living in harsh environmental conditions.

Screening of Awassi Ewes for Milk Yield: Following the logic outlined above, FAO initiated a preliminary programme to test the efficacy of genetic screening (GS) in conjunction with the establishment of Open Nucleus Breeding Scheme (GS/ONBS), as stage one in the genetic improvement of livestock breeds in developing countries. The initial programme was targeted at the genetic improvement of Awassi Sheep for milk production. Pilot projects were initiated in Turkey, Syria and Jordan. In each project, the aim was to screen as many ewes in the national population as realistically possible; the screening was based on subjective assessment (based on interviewing the flock owner) followed by selective validation (measuring the milk yield of ewes claimed to be exceptional in controlled comparison with randomly selected contemporaries).

Table. Genetic Screening Results - Turkey

	Nucleus			**Control**			%
Lactation Yield (kg)p (range)	310		7.7	223		9.3	+39
	(254	-	469)	(97	-	360)	
Lactation Length (days) p(range)	206		1.7	187		4.4	+10
	(159	-	224)	(95	-	222)	
Maximum Daily Yield p range)	2.7		0.1	2.1		0.1	+29
	(2.1	-	4.2)	(1.0	-	4.3)	
No. of Animals	43			43			

Following this protocol, approximately 100,000 Awassi ewes were screened in Turkey, in a cooperative project with the University of Çukurova. Based on on-farm validation (two successive milk yield records), a total of 43 ewes were purchased and recorded in their following lactation alongside a control flock purchased (at random) from the screened population. The results, show that the screened animals outmilked the controls by nearly 40%. This initial response provides very encouraging, albeit tentative evidence that genetic screening for milk yield may point the way forward in the genetic improvement of Awassi sheep; the screening is being continued each year to build up an Open Nucleus Flock, which will ultimately provide improver rams for the industry. Similar screening projects aimed at improving body weight in Djallonke sheep in West Africa (The Gambia, Guinea) have been initiated.

Genetic Improvement Capacity

Very often development planners recommend a breed development programme, which, while technically justified, is without adequate emphasis on and evaluation of the capacity and commitment of the different 'actors' in the breed development chain. Two issues are paramount here: that livestock breed development is a long-term problem requiring, at a minimum, a 20-year commitment; and that breed improvement cannot have any meaningful relevance if it does not compensate the farmer and the other participants in the breed improvement process over a sustainable timeframe and, to this end, research personnel, extension staff and support services (e.g. AI) must play strong and consistent roles to support effective breed development and farmer participation.

Further, long-term government policy and its impact on the development environment will also condition breed improvement possibilities. Consistency and coherence across all the areas in the breed improvement process, broadly determine genetic improvement capacity; all links in the chain are important to the final outcome.

LAND USE, FEED RESOURCES AND INTEGRATED CROP-LIVESTOCK SYSTEMS

The livestock feed resource base within any given land-use farming pattern determines animal productivity; conversely, the land-use pattern sets bounds on the role(s) of livestock within the farming system. The challenge to livestock development planning is to strike acceptable and sustainable balances between these sometimes conflicting interests. The diversity of issues, edaphic, biological, socioeconomic and cultural, that ultimately determine the accepted (traditional) farming system in any given region is very great and far beyond the scope of this chapter.

Only a few selected comments will be made here. Given the agroecological diversity of animal feed resources across the developing world, coupled with the variety of projects and past efforts to improve feed resources, the most that will be attempted is to discuss some general strategies on which livestock feed resources development and utilization might be more effectively based.

Land use

The still rapidly expanding human population in most of the developing world, which is projected to increase from 4 to 7 billion by 2025, continues to focus attention on land use, not only to meet the populations' staple food but also their living space requirements. The trends to urbanization across sub-Saharan Africa, West Asia/North Africa, Asia and Latin America suggest that 54, 75, 56 and 84%, respectively, of the populations on these continents will live in cities by the year 2025 (Winrock, 1992). Both of these fundamental demographic forces demand the attention of livestock planners now and towards the future.

They dictate and emphasis on land-use efficiency in which the synergisms of soil, plant and animal interactions must be fully exploited. Coupled to this, concerns to conserve the natural resource base and to limit the emissions of carbon dioxide, methane and other environmentally damaging impacts, will further add to the complexities of livestock development planning. Bioenergetic efficiency must become the keystone of livestock farming systems in the future. Inevitably, this means integrated crop-livestock systems throughout the tropics and subtropics. Certainly, in the low rainfall (<200 mm/year) rangelands, or in the high altitude alpine pastoral systems, crop livestock interactions will not be strongly evident, but in these environments avoidance of rangeland degradation through overstocking will be of paramount importance. Quite distinct but ultimately compatible short and long-term strategies are required.

Traditional feed resources: In the short term, the development of feed resources to raise ruminant livestock productivity throughout most of the developing world must be based on better use of traditional feeds. Generalizations are difficult to make across the diversity of production systems and their very different feed crops. However, the first step in all systems must be to tackle critical nutritional deficiencies, be they mineral-, protein- or energy-limiting. In extremely difficult conditions, this may simply mean strategic supplementation of the diets of selected poorly thriving animals in order to reduce mortality.

At another stage it will mean improving feed utilization through: the treatment of by-product feeds (ammonia, urea); diet supplementation with balanced highenergy feeds, (urea/molasses blocks, etc.), coupled with the feeding of by-pass protein. The underlying strategy being to build on the existing system and introduce simple and practical technologies to suit local conditions. Parallel to improvement in the feeding regime, due attention must be given to the major prevailing animal disease challenges. In the first instance, this will usually call for parasite control (internal and external parasites) coupled with vaccination against specific diseases as necessary.

New Feed Technologies: In the longer term, adequately supported by ongoing and future research, greater emphasis will need to be focused on the development and utilization of high biomass feed resources (e.g. sugar cane, cassava) for monogastric and ruminant livestock, coupled with better exploitation of high protein forage trees. Expansion of livestock production in the tropics in the future will demand technologies that enable the ruminant to better utilize high fibre feeds, e.g. rumen flora manipulation to improve digestion of cellulose.

Coupled with these changes, livestock planners in the future are likely to have to place much greater emphasis on multipurpose farming systems in which integrated livestock, food and tree crop production maximizes photosynthetic capture to produce food (plant and animal), fibre, energy and fuelwood while maintaining soil fertility and overall sustainability within the system. Examples of such integrated systems are already being developed to exploit the high biomass potential of the tropics. In other areas of the tropics, the more extensive exploitation of integrated rice, fish, azolla and livestock production is being researched (Mukherjee, 1992) in order to develop bioenergetically-efficient production systems to meet food needs in land-scarce countries such as China and South Asia.

THE DEVELOPMENT FORCE

Technical strategies for the advancement of animal agriculture constitute just one vector in the overall matrix of effects that determine the ultimate impact and sustainability of development effort. To place them in their proper context, it is useful to represent the development planning process as an algebraic function of the major determinant vectors. For sustainable livestock development, these vectors will encompass: market demand for animal product or use, a_i; market/animal use channels and pricing arrangements, b_i; land and animal resource potentials and interactions, c_i; practical technical interventions and transfer impacts, d_i; natural resource sustainability and environmental impacts, e_i; and the major human components of the development process, f_i.

This last category ranges from farmer initiative and welfare to the efficiencies of the major government and private support service personnel involved in research, training, extension, market regulation, provision of capital and the formulation and implementation of government policy; these parameters of human endeavour might be conveniently termed the 'development force'.

The challenge, inherent in the development of strategies for the sustainable advancement of animal agriculture, is to fully comprehend the direct (additive) and interactive (non-additive) effects of these different determinant vectors. They are represented in algebraic terms in the following equation, although it is fully understood that it is not possible to emperically formulate such an equation.

$$SLD = f \; B_1 \begin{bmatrix} a_1 \\ a_2 \\ \cdot \\ \cdot \\ \cdot \\ a_n \end{bmatrix}, B_2 \begin{bmatrix} b_1 \\ b_2 \\ \cdot \\ \cdot \\ \cdot \\ b_n \end{bmatrix}, B_3 \begin{bmatrix} c_1 \\ c_2 \\ \cdot \\ \cdot \\ \cdot \\ c_n \end{bmatrix}, B_4 \begin{bmatrix} d_1 \\ d_2 \\ \cdot \\ \cdot \\ \cdot \\ d_n \end{bmatrix}, B_5 \begin{bmatrix} e_1 \\ e_2 \\ \cdot \\ \cdot \\ \cdot \\ e_n \end{bmatrix}, B_6 \begin{bmatrix} f_1 \\ f_2 \\ \cdot \\ \cdot \\ \cdot \\ f_n \end{bmatrix}$$

Modeling and maximum likelihood iterative procedures can help to establish the relative importance (B) of each determinant vector on sustainable livestock development (SLD). Approached in this way, livestock development planning requires an interactive approach in which a number of development scenarios are identified; the interpretation of the outcomes will demand coordinated interdisciplinary analyses involving economists, sociologists, production scientists (land and animal) and environmental impact physics and chemistry. It is beyond the scope of this chapter to critically analyze all the elements of each of the seven determinant vectors identified above. Selective brief comments will be made on two factors in the 'development force' vector in that they relate to issues that are most often ignored.

Farmer Initiative/Welfare: Many development programmes are based on the 'perceived' needs of the target farmers; the word 'perceived' is generally used by planners to indicate a full understanding of what the farmer needs! Farmer needs, like those of any other sector, are a function of tradition, education, and day-to-day survival demands normally influenced by the aspirations of the individuals, the extended family and the local community. Increasingly, farming systems research and extension (FSR/E) approaches are attempting to put real perceptions on these needs; the livestock development planner must follow suit, essentially by 'standing in the farmers shoes' and perceiving development needs and objectives from that position. Concerns for such matters as 'internal rate of return to investment', high output production systems and impacts of new technology will have to be transposed on such vital issues as investment risk, demand on work routines and family welfare since these are influenced by local tradition, culture and status in the local community. The development planning task is to marry these very different perceptions in an acceptable and phased development programme. The roles of the extension and other support services (e.g. credit) have a very important role to play in this reorientation/transition process.

Extension/Research Service Efficiency

Most developing countries, particularly in Africa, dramatically expanded their research and extension services over the past twenty years, basically based on World Bank investment loans. Yet it is difficult to single out many countries that can be deemed to have effective research and extension services to support the development of livestock production. Several International Service for National Agricultural Research (ISNAR) and other studies have identified and articulated many of the reasons why these services have not been successful in supporting change and sustainable advancement. Only two of the many reported constraints on research and extension services in developing countries, discussed in these studies, will be mentioned here; staff motivation and research-extension-support service links. These are often recognized, but are not always given adequate attention by development planners simply because they fall within the realm of national government policy and management capacity.

Staff Motivation: Staff motivation is certainly influenced by training, it can be nurtured or dissipated by management but, in the final analysis, it is heavily dependent on financial reward to the individual. Many research-, extension- or development-oriented projects have experienced great difficulty in harnessing effective commitment from national counterpart staff simply because the staff in question, albeit formally working full time in government service, find it necessary to have a second or sometime a third additional job/activity to earn enough money to support their families. It is unreasonable to accept that development objectives can be realized in such circumstances, although they may be very appropriate in every other way.

Support service efficacy is also heavily dependent on effective professional interaction between public service staff employed by different government departments or agencies. Sadly, it is not uncommon to find situations where apathy, jealousy and sometimes hostility discourage and limit the interactions, exchange of information and cooperation between the different sectors of government support services that are so essential to effective guidance and development of their farmer audiences. The sustainable development of animal agriculture, with its many complex interactions demands that these human initiative issues are fully considered in consort with the other elements in the sustainable development matrix.

SUSTAINABLE DEVELOPMENT OF LIVESTOCK PRODUCTION

The purpose of this chapter is to outline the technical issues and major policy considerations related to the sustainability of livestock development. The identification of these issues is primarily based on the manifestations of unsustainability in certain patterns of livestock production in developing countries. For the purpose of this chapter sustainability of livestock development will be looked at in the context of management and conservation of the natural resource base, i.e. animal and feed resources. The environmental and socio-economic considerations underlying sustainable development are elaborated in FAO (1989).

Livestock development over the past three decades has been mainly directed towards satisfying the rapidly increasing demand for milk, meat and eggs in the urban centres. The required intensification of animal production was often limited by both an inadequate indigenous feed and animal resource base along with a fragile socio-economic capacity in most developing countries. In many cases, the political pressure exerted by the urban demand led to the importation of exotic stock and concentrate feeds. Technologies which proved successful in the socio-economic context of developed countries were applied. Technical institutions were also founded on similar rationale. Large amounts of public funds were used to support the modern production systems. The traditional livestock sector was also stimulated by the increasing prices for animal products but its growth was hampered by inadequate input supply and support services.

The signs of unsustainability of such development began to surface in the past few years. The sudden increase in grazing pressure and lack of capacity for conservation among the graziers led to the degradation of rangelands. Feed imports could not be sustained. Lower adaptability of imported stock did not allow rapid multiplication and crossbreeding raised the fears of erosion of the indigenous genetic diversity. The pollution of water and air has been limited to the few areas of livestock concentration in intensive production units. The large-scale use of insecticides in the past for Tsetse control and sporadic pollution of waterways with acaricides may have had significant environmental consequences.

Apart from the manifestations of unsustainability, the overall increase in animal productivity in the past three development decades has been minimal. Increased external assistance *per se* did not have a lasting impact on the performance of the livestock sector in developing countries. Particularly, smallholder or rural livestock production did not pick up any significant momentum.

DEMAND-DRIVEN DEVELOPMENT AND SUSTAINABILITY

Development Patterns

The high economic demand for milk and meat in developed countries has led to the establishment of capital intensive systems that require high animal productivity, high levels of concentrate (grain) feeding and highly mechanized or automated infrastructure. The production technologies developed over the years have successfully met these requirements. The problems currently faced by these systems arise basically from economic pressures and revolve around capital costs, input costs and subsidies. There are also environmental problems associated with high concentration of stock, animal wastes and high levels of fertilizer use for pastures. Feed availability is not a constraint in feed deficit areas thanks to the well-established trade in feedstuffs. Large quantities of feed are imported from developing countries. In 1988, some 3.7 million tons of cereal brans and oilseed cakes were imported in the developed countries from the feed-deficit developing countries.

The prevailing production systems in the developing countries are a world apart. Large numbers of ruminant livestock graze common grassland especially in Africa. This low-input system had, in the past, successfully met a large part of the demand for meat in urban centres and catered for the local demand for milk and meat. For the past 2–3 decades the system has not been able to meet the increased demand arising from population explosion, increased income and rapid urbanization. Livestock numbers on common grazing areas have increased way beyond the carrying capacity. The lack of economic capacity among the graziers to undertake range improvement resulted in degradation of most of these grasslands. After an increase in numbers and offtake during the past 2–3 decades, the present output is declining. The degraded common grasslands cannot be rehabilitated without massive investment and the politically difficult land tenure policy.

A sizeable proportion of the third world livestock is also kept in arable farming areas and is integrated in the system to provide draught power, fibre and dung as well as marketable quantities of milk, meat and eggs. A few holdings in these areas are large enough to warrant increasing the producing ability of the stock. On small farms, the large ruminants are kept primarily for work; milk and meat are by-products. Small ruminants are primarily kept for meat; milk is a by-product. Crop residues are the main feed resources although a small part of the land is devoted to fodder for work animals. Poultry and pigs basically scavenge the farm and household wastes. Supplementary feeding depends on the market for produce. In recent years, the increased price of milk and meat and improved marketing opportunities for the small farmers have, in some cases, resulted in increased fodder cultivation, better utilization of crop residues and by-products, supplementary feeding and animal husbandry geared towards marketable milk, meat and egg production. However, except for poultry, these improvements have not taken root to a significant extent.

Spurred by high demand for milk, meat and eggs in urban centres, heavy investments were made in dairy farms, ranches and feedlots following the developed country model. This was facilitated during the mid-seventies to mid-eighties by growth in external loans, foreign exchange earnings and availability of capital at low interest rates. In most countries which encouraged this model the available foreign exchange was used to import high yielding stock and feedstuffs. However, the contribution of these enterprises to total milk and meat supply has been minimal and at high cost. With increased capital and foreign exchange shortages in recent years, the specialized livestock farming enterprises and feed industries are trying to re-orient their operations towards greater reliance on locally available feedstuffs and animal resources (FAO, 1990a). There is a growing realization of the value of multipurpose local stock and their capacity to utilise fibrous feedstuffs.

A major part of milk and meat in developing countries is produced by the resource-poor small farmers. Smallholders contribute by far the largest part of the labour force. To improve production and income, techno-economic changes must be brought in the small farming system. Availability of sufficient feeds of adequate quality is the basic constraint in the system. Possibilities of providing external feed inputs are limited. Due to a lack of sufficient attention to the smallholder, appropriate technologies to improve performance of locally available animal and feed resources within the rural system have not developed. Institutional capacities in developing countries have also not been built up from this viewpoint. Smallholder dairy development, for example, has succeeded only when appropriate production technology was supported by an integrated dairy development programme incorporating milk collection, input supply and favourable price policy (FAO, 1984).

Technology Requirements

The objective of livestock development at the national level has been to attain as much self-sufficiency as possible to satisfy mainly the urban demand. At the farm level, the objective is to increase income and utilise family labour year-round. The role of animal production technology in developing countries revolves around finding labour-intensive procedures which maximise production with low-cost inputs. Feeds comprise by far the largest component of input cost. The efficiency of feed utilisation in terms of herd or flock output is the primary consideration in technology development. Other elements of the development of appropriate and affordable technologies that ensure sustainability are: conservation and improvement of resource base; minimization of wastes and environmental degradation; and recycling of wastes for animal feed or biogas.

Direct transfer or transplantation of technology from developed to developing countries has rarely been successful and sustainable in achieving the above objectives. However, in the case of a few specialized enterprises where such a transfer has been successful, the longer term consequences have been dependent upon imported feeds and exotic animals in order to take maximum advantage of the transferred systems. The side effects of imported technologies has been the neglect of indigenous livestock and feed resources.

A sustained growth of indigenous production of milk and meat is dependent upon the introduction of new technologies that would be adopted by producers, especially, the smallholders. A systems approch is indispensable for the development of such technologies. This approach should incorporate the rigour of the scientific method as well as the human element of involving the producer of the farmer at all stages of development. There is a pressing need to utilise the effective tools and techniques that are available to pursue a farming systems research and extension strategy. The farmer must be a key partner in every effort in the improvisation and introduction of the required technologies.

Multi-disciplinary effort and effective linking of research, extension and training are important requisites for the systems approach. The process of increasing animal productivity must also be effectively linked with other developmental elements such as market considerations, produce organization, incentives and policy-making. Vigorous research-extension efforts have proved to be the basic requirements for improving animal production systems.

During the past three decades, most developing countries have been able to establish institutional infrastructure for livestock development, e.g. research centres, extension services, veterinary laboratories, disease control services and educational institutions at various levels. The technical performance of this infrastructure is variable from country to country and from institution to institution. On the whole, the institutional impact on livestock production has been open to question. The development concern these days is not so much about the capacities in terms of physical infrastructures or size of trained manpower but about the usefulness of this capacity in improving farm output. The underlying constraints emerge from the prevailing traditional, neo-social, economic and political environment.

Self-reliant livestock production in developing countries requires a strategy to optimise production from available feed resources through an integrated technology which employs multi-purpose crops, multi-purpose animals and recycling of residues and by-products. The identification of needs and a careful study of feed and animal resources are essential first steps. Resources must be examined in the context of various agro-climatic zones and fodder crops which might be grown. Ruminant production systems must then be matched with the resources in a way that aim for economic rather than biological maximisation. To introduce new technologies it is important to start with on-farm improvements. The utilisation of locally available feed resources must be maximised to reduce or eliminate the importation of concentrate feeds (FAO, 1985).

Livestock development efforts in the past laid primary emphasis on rapid genetic improvement arguing that improvements in feeding will be ineffective when animals with low genetic potential are raised. In recent years there is a growing consciousness to balance the rate of genetic improvement with improvements in feed availability and management. There is also an increased realization of the potential of indigenous cattle, buffaloes, sheep and goats as multi-purpose animals suited to sustainable production systems and as efficient convertors of locally available feed resources. Innovated breeding and management procedures have not been effectively implemented to improve reproduction and to increase the milk, meat and work outputs of the indigenous stock.

MAJOR ISSUES

Technology Development

Developing technologies for sustainable animal production has always been the objective of animal scientists even this was not expressly stated. This concern had been well taken into account regarding the environment in which the technologies were developed. However, the use, rather than misuse or misplacement, of a proven technology could be an issue in the context of sustainability in different environments. The important issue in this regard is the development of appropriate and affordable livestock technologies suited to specific agroclimatic zones in developing countries. To satisfy the criteria for sustainability, these technologies should support agricultural development which "conserves land, water, plant and animal genetic resources, and is environmentally non-degrading, economically viable and socially acceptable" (FAO, 1989). This would mean further research and development effort for the expansion of the feed base, conservation of animal genetic resources, recycling of wastes and efficient feeding systems.

Methodology to evolve sustainable production systems requires a multi-disciplinary approach and development of appropriate indicators of medium-term or long-term sustainability. These indicators should be practical enough to be used in project formulation, monitoring and evaluation.

Sustainable Institutional Support

The institutional framework for improving production systems needs to be structured, or restructured, to ensure multi-disciplinary collaboration and cost-effectiveness. Participation and cost-sharing by the farmers' organizations may be necessary in most cases. Present institutional structures for research, extension and veterinary services need evaluation.

Import/Export and Price Policies

Past experience has highlighted the unsustainability of livestock production and feed imports in most developing countries. The impact of imports on the existing production systems needs to be fully understood before allowing the import of milk, meat, feeds or live animals. Similarly, the export of agro-industrial by-products from feed-deficit countries disrupts livestock development efforts.

Price policies for milk, meat, feed or eggs should be geared towards creation of an economic environment in which it would be profitable to conserve and utilise local feed and animal resources for production. These policies should be devised and implemented for the benefit of the producer.

Degradation of Grazing Lands

Common grazing lands are most vulnerable to degradation. The basic issue in rangeland rehabilitation is land tenure policy which would encourage range improvement and proper grazing management. Access to the rangelands by the nomads and other poor segments of the population is also an issue.

Conservation of Indigenous Breeds

The relevance of the conservation and genetic improvement of indigenous animal genetic resources is well recognised (FAO, 1990b). The important issues concern the cost and the cost-effectiveness of methods to be employed. It has often been mentioned that these efforts should be supported by public funds and external assistance. However, this point of view should not divert the attention from efforts to develop privately-supported and cost-effective methods for conservation.

Methane Emission by Ruminants

There are various estimates of the contribution of ruminants to methane production and global warming. The reliability of these estimates apart, the issue concerns the options available to reduce methane emission. One option on which greater concentration of efforts may be needed is that of devising practical feeding procedures to reduce methane production per unit of milk or meat output.

INVESTMENT FOR SUSTAINABLE LIVESTOCK DEVELOPMENT IN DEVELOPING COUNTRIES

Bank livestock lending peaked in the 6-year period 1974-79 when an annual average of seven free standing livestock projects and 19 projects with a livestock component were approved; costs including the cost of livestock components amounted to US$9.47 billion (1989 US dollars) and 5.15 billion (U.S.$ current). Livestock lending declined in the period 1980-85 to a yearly average of about two free-standing projects and 17 component projects; costs amounted to 5.77 billion (1989 US dollars).

Over the five year period 1986-90 the aggregate cost of Bank-assisted livestock projects and components amounted to US$3.54 billion (1989 US dollars). Average annual lending declined in real terms from US$1,578 billion for period 74–79 to US$962 million for period 1980 to 85 and US$708 million for period 1986–90.

This means that average lending for the last two periods was only 61% and 45% of the earlier period (1974–79) in real terms.

For the first time no free standing livestock project was approved in FY90 but 11 projects were approved with a livestock component. The absence of a free standing project in FY90 is considered as a one-year anomaly and it is expected that the number of livestock projects per year will stabilize at about 2 free-standing and 12–15 component projects.

The reduction in livestock lending as a proportion of the total lending for agriculture parallels, albeit to a greater extent, agriculture's decline as a percentage of the Bank's lending operations; from 30.1% in 1980 to 16.3% in 1989 and 17.3% in 1990. The character of lending operations changed substantially over the last decade with greater emphasis being placed on lending instruments other than specific investment loans. IBRD and IDA lending by loan category is given below for FY90.

Loan Category	US$Million	%
Specific investment	19,179.9	49.15
Sector investment	3,957.2	19.11
Financial intermediary	879.0	4.24
Sector adjustment	2,543.6	12.28
Program lending and Structural adjustment	1,434.0	6.92
Debt reduction	1,460.0	7.05
Technical assistance	203.0	0.98
Emergency reconstruction	54.0	0.26
	20,710.7	100.00

A decade ago Bank lending was dominated by specific investment loans but the picture has changed drastically with the increased emphasis on other types of loans, especially sector investment loans, sector adjustment loans, program lending and structural adjustment loans.

BANK EXPERIENCE WITH LIVESTOCK LENDING

The reduction in Bank lending for livestock was caused by the relatively poor performance of livestock projects as indicated by a number of reports and papers which reviewed lending for livestock development. The most important and comprehensive of these reviews- "The Smallholder Dimension of Livestock (1985)" - was undertaken by the Bank's Operations Evaluation Department (OED) which is entrusted, by the Bank's Board of Directors, with authority and responsibility to undertake an independent review and report on the performance of all Bank lending operations. In addition to the OED report the Bank's experience with Dairy Development was reviewed in 1982 (AGR Technical Note No. 6).

The Bank's experience in Dry Tropical Africa was reviewed in 1981 by Mr. Stephen Sandford (Consultant) who produced a report for internal use. These three reports provide a good independent assessment of livestock lending including problems and issues and suggested lessons. Four World Bank reports which dealt with special aspect of livestock, e.g., veterinary, dairying and integrated crop- livestock are not dealt with in this chapter although they provide valuable insights on livestock lending and the sustainability of development efforts.

OED Report

The OED report was based on a review of 124 audited projects and 206 ongoing projects which comprised the Bank's livestock portfolio at the end of 1983. Of the 330 total, 91 were livestock projects and 239 livestock component projects. Of the 124 audited projects, 52 were livestock projects and 76 were component projects.

From modest beginnings in 1959 in Uruguay, through late 1983 early 1984 when the OED study commenced, the Bank provided some US$11.7 billion (in constant 1983 dollars) for livestock development of which almost US$6.1 billion (52%) was targeted to smallholders. The livestock sub-sector was thus significant in the Bank's lending portfolio and smallholder livestock lending was an important part.

The rate of livestock lending increased through the 60's and 70's to peak in 1979 and thereafter to decline. Smallholder lending followed a similar pattern but the percentage of total livestock investments for smallholder development showed a steady increase from the late 60's to the present and over the 70's accounted for roughly two-thirds of all Bank livestock lending. This was in response to Bank management's increasing sensitivity in that period to equity considerations which affected all agricultural sub-sectors.

There were according to the OED report regional differences in project components and their design, in species/product emphasis and in target group, depending on ecological, socio-economic, cultural and traditional factors. Credit and livestock purchase were major project components in most regions. Nearly two-thirds of all audited projects included such components. About half of all projects included components for development of on-farm infrastructure, pasture improvement, and fodder development. Projects frequently included the above activities as multiple components. About 40% included a technical assistance and/or farmer training component. Other components occurred less frequently. In general, the design of smallholder projects was similar to that found for the whole set of livestock-related projects. The ongoing projects also had larger scope than the audited projects and nearly all of the components appeared with greater frequency suggesting that each project, on average, included a larger number of livestock-related components.

In terms of species emphasis, the OED report found that investments in cattle development (beef, dairy, and dairy-beef, in that order) accounted for about two-thirds of component activity in audited projects and an even higher proportion of total funds. Substantially less attention was paid to other species, viz. sheep, poultry, and swine, each about 10%; goats, about 5%; and miscellaneous species (cameloids, rabbits, bees, etc.), about 2%. Cattle development was also emphasized in ongoing projects, but its relative importance decreased to less than one-half of the total number of interventions. Thus, while cattle development continues to be important in ongoing projects, other species are now increasingly emphasized. This emphasis on other species reflects increased efforts to support and improve existing smallholder farming systems. In such systems there is also growing emphasis on integrating livestock with agriculture, to their mutual benefit.

All regions participated in the Bank's livestock development efforts according to the OED report. The number of livestock-related projects (audited and ongoing) at that time was spread fairly evenly by region, but the bulk (three fourths) of all livestock investments were in Europe, Middle East and North Africa Region (EMENA) and Latin America and the Caribbean Region (LAC). Smallholders received the highest proportion of livestock investments in Western Africa Region (WA), South Asia Region (SA), EMENA, and East Asia and Pacific Region (EAP) in audited projects, and in Eastern and Southern Africa Region (ESA), EAP, SA, and WA in ongoing projects.

Project Performance

The average ERR of all audited livestock activities Bank-wide was 11% (OED report). The average ERR of 46 livestock projects was 7.2% and that of 58 component projects, 14%. The lower average figure for livestock projects appeared to be due to the low returns from smallholder livestock projects (average ERR -0.3%) and large holder livestock projects (average ERR 6.2%), as mixed smallholder/largeholder livestock projects performed satisfactorily on average (ERR 11.6%). Smallholder component projects (average ERR 10.7%), largeholder component projects (average ERR 18.9%) also performed generally satisfactorily. While the results suggested that livestock investments made as a component of a diversified project were most successful than as a part of a straight livestock project, there was insufficient data to support such a conclusion statistically because separate ERRs were rarely available for individual components of multi-component projects, particularly if the individual components were relatively small, as they often were for livestock investments. It was stated that additional information was needed on the performance of livestock investments within livestock component projects, particularly as such investments comprised an increasing proportion of the total ongoing livestock portfolio at the time of the OED Study.

By region the OED Report showed that, EMENA, LAC and Southeast Asia had the highest number of total projects with ERRs exceeding 10% (77%, 68% and 67%, respectively) and ESA, EAP and WA had the highest number below

10% (74%, 56% and 50%, respectively). Some 14 of 22 projects (64%) with negative ERRs were in ESA and WA. By project type, livestock projects performed particularly poorly on average in both African regions and, to a lesser extent, in East Asia and the Pacific Region (EAP). Component projects also performed unsatisfactorily on average in Eastern and Southern Africa (ESA), but performed satisfactorily overall in all other regions, particularly in EAP which had the highest average ERR (33.7%) Bank-wide.

Ex-ante appraisal projections were observed to be much more optimistic than *ex-port* ERRs. *Ex-ante* ERRs on 46 livestock projects were some 200% higher than ex-post ERRs and on 58 component projects they were some 70% higher. Only one in six of all projects has ERRs at completion equal to or greater than appraisal ERRs.

Two conclusions emerged from the OED Study. First, a large number of livestock investments were successful, particularly in the regions where lending was highest, and their success should not be obscured by the existence of problem projects, particularly those in ESA and WA. Second, the substantial variation in project performance suggested a need for improved appraisal methods, especially greater attention to the production coefficients adopted, the benefit stream projected, the project time frame and risk analysis.

The principal factors affecting project outcome identified in the OED study were the availability or lack of:

- Technological packages adequately adapted to existing farming systems;
- An economic context providing attractive producer incentives;
- The institutional capability for implementing the proposed project;
- Qualified technical personnel;
- A government commitment to livestock development and/or smallholders;
- Political and economic stability;
- Clear property rights for lands to be developed;
- Functioning producer organizations - particularly where group action is needed;
- A project design which realistically takes into account country strengths and weaknesses; and
- Firm, consistent, but flexible supervision of implementation.

The OED report stressed that these factors were similar to those which cause problems in projects in other agricultural sub-sectors. An effort was made to identify factors specific to livestock. The risk in livestock projects appeared most closely linked to the inadequacy of applied livestock-related research in most countries, the lack of technical personnel, the weakness of livestock-related institutions and their lack of integration with agricultural institutions, the greater importance of land tenure issues, and the failure of some projects to recognize fully the inadequacy of the "base" on which projects had to build. The tendency to proceed too rapidly in terms of physical implementation, without sufficient technological, institutional and staff development, stood out.

The study also pointed out that the performance of livestock projects must also be measured in dimensions other than the simple ERRs. Many livestock projects were pioneering efforts, involving new relatively untested technologies, requiring institutional strengthening, staff development, and livestock policy formulation. Benefits achieved through improved institutions, staff development, and "learning by doing" were not revealed in the ERRs. Nonetheless, there was a strong correlation between the project ERRs and the audit reports assessments of achievements in these other areas - poor economic performance has often been accompanied by poor institutional development and the like rather than offset by improvements therein. Indeed, poor performance in these areas was a major cause of low ERRs.

The smallholder livestock and component projects examined by the OED Study fared worse than livestock projects (taken as a group). Their average *ex-ante* appraisal ERR was equally as high as other livestock projects, but their *ex-post* ERR was even lower.

The OED study concluded that the widely-held perception that the Bank's livestock development efforts have been unsatisfactory accounted, in part, for the steep decline in livestock lending since 1980.

The study found that the performance of Bank livestock was highly variable, ranging from very satisfactory to very unsatisfactory, but was satisfactory more frequently than not. It also appeared that considerable learning had taken place regarding the design and implementation of livestock projects, and it was concluded that future projects should perform better. Nonetheless, it also concluded that the evidence suggested that livestock projects overall may be more difficult than other agricultural sub-sector projects, and that livestock assistance, especially to smallholders, would probably require greater design, implementation, and supervision inputs than they had received up to that time. The study also concluded that the Bank should continue and probably increase its support for livestock development, especially to smallholders, given the high potential for raising their incomes and living standards. It pointed out that livestock were a key element in raising farm productivity and it was difficult to conceive of sustained increases especially in smallholder agriculture in most areas in the world without attention to livestock development; demand for livestock products was increasing rapidly in most developing regions, and livestock investments were expected to be increasingly economically attractive.

The study highlighted the point that livestock project design initially placed heavy emphasis on meat and milk and largely ignored other outputs such as traction and manure. This approach was strongly influenced by livestock systems in developed countries, and by an emphasis on larger commercial producers. The move toward smallholder livestock development had encouraged a shift toward more diversified use of livestock and the integration of livestock and agricultural production activities. It emphasized the need for this shift to be more fully reflected in project design, e.g., greater cognizance should be taken of joint livestock outputs when assessing the demand for and the benefits of livestock production, and greater effort should be made to coordinate the efforts of livestock and agricultural development agencies.

Milk was considered to merit greater emphasis relative to meat production. The primary need was seen as the organization of marketing, processing and distribution facilities, especially in regions where milk production was dominated by smallholders. Small-scale dairying was seen as undoubtedly one of the most promising avenues for future Bank lending.

It concluded that livestock development efforts suffered from a "cattle bias". Additional emphasis should be placed on swine and poultry, small ruminants and other animals, especially through research, technical services, and market development.

The study made special mention of livestock-related projects in the African regions because of the difficulties which were experienced there. It pointed out that many of the countries in ESA and WA were newly independent with ill-defined policies and priorities, limited infrastructure, weak skilled manpower and material resources, widespread poverty, and a high prevalence of drought and pestilence. Many governments were overly centralized and urban oriented. Not only livestock investments fared poorly in such countries, but indeed all agricultural-related activities. Nevertheless, in several countries, livestock development appeared crucial to overall economic development and, in others, it would have a high positive impact. It emphasized that a substantial amount had been learned regarding livestock development, and there was evidence that where such lessons had been applied in ongoing projects the situation was improving.

Finally, it was evident from the study that smallholder livestock projects performed unsatisfactorily overall. The review indicates that efforts were often made to develop individual projects which were innovative, but these were also often ambitious in scope and size, were generally weak technically, and were implemented in a largely unfavourable economic climate in countries where government sometimes showed limited sensitivity to smallholder development potential and needs. Target groups sometimes failed to involve themselves in project design and implementation out of misunderstandings or from distrust of government intentions, and were other times excluded either for paternalistic or political reasons. The Bank may have been too ready to finance such ambitious projects, particularly where institutional support was weak, where land tenure problems were apparent, and where government commitment was questionable. In a number of instances, an exploratory pilot phase would have been more appropriate instead of a large, demanding and high-risk effort. The study mentioned that a number of operational staff hold the view that pressures of the lending program were a contributing factor in this context.

BANK-FINANCED DAIRY PROJECTS

In 1982 the Bank prepared a comprehensive review of Bank/IDA-financed Dairy Projects (AGR Technical Note No. 6). The paper lists and reviews 75 projects which were either exclusively for dairying or had a dairy component. The total cost of these projects amounted to US$6,533 millions, the dairy components amounted to US$1,034 million and the loan/credit amounted to US$1,999 million.

Latin America (LAC) Dairy Projects

The study concluded from a review of 30 dairy programs in LAC that the projects "successfully attained production targets, improved the institutional support structure and contributed towards the establishment of effective livestock credit systems".

A feature of the region was the traditional preference for raw milk (which is boiled before use) and this enabled smaller producers to dispose of surplus milk directly to consumers without concern for an accessible processing facility. Low priced imports were a constant threat but it is concluded that "the livestock industry throughout the area had adjusted to the low price import option by developing a dual purpose production system, by utilizing natural pastures, by upgrading cattle and by increasing the carrying capacity of the land". Project officers considered that marketing patterns did not need changing and processing facilities, especially for pasteurized milk, were adequate although increased investments would be required as dairying expanded. The report draws attention to the deleterious effect of low cost imported milk products on investment and production in many LAC countries but concludes "in the LAC area it is doubtful that there is a country or group of countries which could be described as marginal for dairy production in terms of the possible financial advantages of imported milk projects".

Europe, Middle East and North African (EMENA) Dairy Projects

Fifteen dairy or projects with dairy components were assisted in six countries (Morocco, Turkey, Yugoslavia, Ireland, Spain and Romania). Total project costs were US$2.47 billion and dairy development components amounted to US$778.01 million and Bank loans amounted to US$597 million. Projects are judged to have performed satisfactorily in all cases with the exception of large public sector farms in Yugoslavia. They encountered serious management, overstaffing and price problems. Yugoslavia changed its policies as a result of this negative experience and subsequent projects supported smallholders only.

Dairy programs in virtually all countries emphasized pasture, forage and feed production. Successful dairy cattle importations were a feature in the case of Turkey and Morocco (Friesian and Brown Swiss). Another feature of these latter countries is the prevalence of raw milk consumption which enables producers to sell directly to consumers and traders without incurring processing costs. The Rumanian projects main objective was to improve yields and labour productivity by modernizing existing large public sector units (involving feeding, housing and milking). With the exception of Ireland, which exported most of its milk as manufactured products under EEC arrangements, production in the EMENA projects was almost entirely for home consumption to meet increasing consumer demand.

The report concludes: "Assessment of overall performance of EMENA dairy development projects provides evidence among most projects of measurable success in achieving project goals and governments' major objectives. The diversity of project design grew out of the need to develop approaches best suited for particular social, economic, and ecological conditions country by country. Successive programs in Turkey, Morocco and Yugoslavia built on earlier experiences, revising where necessary, and in the case of Yugoslavia, making a major shift from large social sector production units to emphasis on smallholder production. This shift of emphasis had also occurred in Turkey and Morocco and represented a welcome development which appeared to be the trend in all regions. The Rumanian experience represented an isolated set of conditions and as noted above was probably limited to one country".

Eastern Africa Region (EA) Dairy Projects

Bank-assisted dairy development in EA, six projects in four countries (Ethiopia, Kenya, Tanzania and Zambia) and a modest component in a Malawi rural development project, was small. Total project costs amounted to US$88.5 million, total dairy components amounted to US$65.4 and IDA credits amounted to US$56.6 million. The Ethiopian project was greatly affected by major political changes and was generally unsuccessful although the country has a good potential for dairying.

The dairy components of three Kenya smallholder credit projects were highly successful. The extent of success is indicated by a consumption level of 75 litres/capita, by smallholders supplying 75% of the total market supply. The Tanzania project supported large units (350 cow units) under parastatal management and since 13 out of the proposed 17 units established did not cover operating costs, because costs were high and production coefficients were much lower than appraised estimates, the project was a failure.

Although a smallholder component was included in the Tanzania project, it was seriously constrained by a shortage of grade cows. The Zambia project was drastically revised downwards after a review and the revised project involving 150 smallholders instead of the original 1,800 performed satisfactorily.

South Asia (AS) Dairy Projects

Total investment in dairy development has amounted to US$547 million, total project costs amounted to US$556 million and loans and credits amounted to US$250. Four dairy projects in India and one in Sri Lanka were supported. In addition two livestock projects in Burma and Pakistan had dairy components. The Indian dairy projects have been singularly successful. Dairying in India is characterized by the development of well-managed cooperatives which handle collection, processing and marketing, and provide support services efficiently to existing small dairy farmers, who typically own one or two milking buffalo. Although production conditions in Pakistan are similar but generally superior to those in India and although the project aimed at replicating the Indian Amul model as far as possible, its performance has fallen far short because the project was not as well managed and was not as successful at institution building, at commanding government support or at defining and implementing appropriate pricing and marketing policies. The Sri Lanka project was drastically revised after initial disappointing experience to replicate the Indian/Amul model as far as possible. After revision, the performance was fairly satisfactory. A project feature was the poor performance and high mortality of heifers imported from Australia. The Burma project had limited success because the Socialist government showed little interest in supporting and developing the smallholder dairy sector despite its considerable potential.

East Asia Pacific Region (AE) Dairy Projects

The regions total investment in dairy production was US$50.3 million, total project costs were US$122.6 million land Bank loans amounted to US$62 million. Two dairy projects were assisted in Korea. A smallholder coconut development in Malaysia had a fairly substantial dairy component costing US$13 million and under the Philippines Second Livestock Project a small pilot dairy component costing US$0.225 million was supported. The Korean projects were highly successful in financial terms and production coefficients in most cases reached or exceeded appraisal expectations. However, they were judged to have negative economic rates of return by the Bank when opportunity costs of imported products were taken into consideration.

Under the dairy component of the Malaysian project the importation of 6,600 heifers for Government raising centres was envisaged. Serious problems were encountered with imported heifers, pasture development was much slower than expected and government tended to support large scale public and private enterprises over smallholders. A Bank supervision mission recalculated the ERR and showed that it was negative. It was shown that milk, reconstituted from imported ingredients, cost US$0.22/litre compared with US$0.24/lither for local milk delivered to the processing plant. On the basis of this analysis, government was advised to slow dairy development and most of the available funds were not used. Malaysia has an extremely humid tropical climate and Bank staff consider that the potential for dairy development is extremely limited. The Philippines pilot dairy component had limited success. Although the Philippines has some potential for dairy development despite the humid tropical climate, Bank staff are of the opinion that dairy development there will be slow.

WORLD BANK LIVESTOCK ACTIVITIES IN DRY TROPICAL AFRICA

The 1981 "Review of World Bank Livestock Activities in Dry Tropical Africa" covered 34 livestock and 37 mixed livestock/crop projects in 26 Sub-Saharan countries. It was undertaken by Mr. Stephen Sandford who had considerable experience of livestock development in Africa. The review dealt primarily with 30 "livestock only" projects (in 22 different countries) and only cursory reference was made to 37 mixed projects. Although the review was based on information in Bank Documents, the author drew heavily on his independent knowledge and broad experience of livestock production and pastoralist in arid and semi-arid regions.

The report concluded that "ranching" projects or ranching components failed dismally. It also concluded that components to improve marketing and livestock movement, slaughtering and processing "have an abysmal record" and that veterinary components and off-range fattening by smallholders have a "generally good record". Sandford was unable to show any statistical relationship between failure and the size and complexity of projects, but nevertheless concluded "... it is my belief that projects are too big, too complex and excessively dependent on expatriates". He further concluded "... that the increasing size of projects during the 1970's was more related to the Bank's own desire to spend more on agricultural sectors than on a realistic assessment of the absorptive capacity of the livestock sub-sector". He also argued that the Bank emphasizes what happens at the top — on the performance of the bureaucracy — and underemphasizes what happens at the bottom — actual performance at the field level.

Role of the Bank

In Sandford's view, the Bank, apart from providing capital, provides three other important elements: (a) pressure on governments not to neglect their livestock sectors; (b) specific pressure in favour of particular policies, programs and components; and (c) technical advice on particular points. Overall he felt that little of the increased meat and milk production during the previous 20 years could be attributed to Bank projects or government programs if rinderpest measures and water development were excluded. He considered that what development there was came from a growth in livestock populations, required and made possible from growth in the human population and some expansion in livestock forage (crop residues). Despite this, he cautioned that livestock production and the welfare of livestock owners will decline further "... unless more effective and more wide-scale government sponsored programs are undertaken" because of the decreased availability of land for extensive production systems and the encroachment of cultivators on grazing land.

Despite his criticisms, he believed that the effect of the Bank's involvement on livestock development had "been beneficial" although over-influenced by "fashions" such as ranching in the '60s, fattening in the early '70's and group formation more recently. He concluded "... Bank-financed projects are usually better oriented, as well as better financed, than other livestock programs implemented by governments, and the visits of Bank appraisal and sub-sector review missions are often the occasion on which government think most deeply about their livestock policies and programs". Although he concluded that "... Bank expenditures (US$750 million) on livestock development will not be justified by production increases in the short term, important lessons can be learned from the experience" and management cadres in African countries are being slowly build up". Livestock programs in Africa should be less capital-intensive, smaller amounts should be spent more slowly and much greater flexibility should be permitted. From the viewpoint of African welfare and economic growth he felt that it would be a pity "... if the Bank were to conclude that if it cannot spend very large sums fast, then it has no proper role in the future in African livestock development".

Rangelands

Sandford feels that there is little reliable evidence "...indicating either that widespread degradation is going on in African rangelands or that, if it is, it is due to livestock development programs". However, he felt that continued caution over the development of water supplies was warranted, but as much for social as for environmental reasons. He considered the Bank's increased willingness to finance veterinary components favourably because of the positive effects on the welfare of poor and rich stock owners and because he does not accept the environmental danger argument. He condemns "ranching" as opposed to pastoralist systems on many grounds including:

- Ranches are less equitable because they increasingly favour the rich and powerful;
- They are often a reason for expropriating land which is already being fully or partially used;
- At least, under African conditions, ranches have no advantage in terms of increased food production over pastoralist systems or smallholder herds;

- At least in terms of production coefficients there is no reason to favour one type of ranch (pastoral, cooperative, company group or private) over any other commercial arrangement by pastoralist; and
- There is little evidence that animal numbers (stocking rate) can be controlled better on ranches, including private ranches than under pastoralist systems.

Sandford does not agree despite many claims to the contrary "that the know-how already exists to improve range and pasture productivity in African arid regions (less than 600 mm)". Although he appears to agree that technology exists to improve rangelands and productivity in the higher rainfall areas (over 600 mm) major questions remain to be resolved, e.g., is technology cost effective and how can the management capability needed to successfully implement it be developed? (In view of the Bank's experience with ranches, and its justifiable reluctance to support them, the question is probably moot at this stage except for smallholders). Sandford placed considerable emphasis on the need for increased support for livestock research and attention was drawn to the smallness of research components in Bank projects (about 1.4% of total project costs). A recommendation was made "that the Bank should allocate a substantially higher proportion of its livestock programs for research" and it should adopt a more active policy of assisting research units in major livestock countries.

In view of the Bank's negative experience with livestock marketing and processing components Sandford recommended a special study on these. Much more attention should, in his view, be given to ranch records and information on ranch performance, in the Bank's records, is judged to be grossly inadequate. Emphasis was placed on the need to make greater use of "competent anthropologists" in project preparation and implementation. Although he refers to the use of anthropologists in 30% of livestock projects, there appears to have been no measurable difference between projects whether or not anthropologists were used. Sandford considered that veterinary components of livestock projects were successful although the extent to which costs were recovered was not clear. Producers put high priority on animal health inputs and the evidence available showed that they were prepared to pay for drugs, vaccines and medicines if given the chance to do so. He recommended that considerably more support should be given to veterinary research to fill the gap in coverage between the research programs of ILCA and ILRAD, to provide guidelines on how veterinary services should be improved and how field delivery systems should be organized and developed.

Pastoral Associations

Although the recent emphasis on pastoral associations (Sandford includes group ranches in this category) was considered "a move in the right direction", he cautions against seeking some "universal model". Since he was sceptical about the value of grazing management or controls on animal numbers, except to the extent that pastoralist may themselves handle these matters, he was against their use as vehicles for implementing government controls on grazing or stocking rates. Association's main functions, in his view, embraced land tenure, resource management, provision of services, communication of information, external relations and the building and maintenance of community cohesion and morale. He made the point that associations have land tenure and land reform implications and they should be treated like land reform projects.

General Problems

Staff-related problems, such as the inability to recruit suitably qualified staff occurred in 90% of the projects analyzed (by Sandford) and political policy-related problems in about 70%. In Eastern and Southern Africa, government policies were one of the most frequently cited problems. Formal project coordination committees worked badly, especially if established at a senior level. Project management costs, although difficult to estimate, appear to have been excessively high in some cases.

The Sandford review takes issue with the project approach towards livestock development in Africa and argues for a program approach. It also argues for flexibility in design and implementation of projects in dry regions to enable management to deal with unforeseen circumstances such as droughts. Sandford believes that the Bank becomes over-involved in detail and that Bank project appraisal methodology "has more to do with Bank ritual than with the effective design of projects".

INVESTMENT IN LIVESTOCK IN DEVELOPING COUNTRIES

Present Status of Livestock Projects

The performance of livestock projects is still a cause for concern. On the Bank's internal rating system, mandated for all projects under supervision, the performance of 40% of livestock projects in 1989 was unsatisfactory. Only Fisheries projects have a worse performance record and even area development projects, although bad, are rated somewhat better than livestock. This rating refers to free standing livestock projects and must be interpreted with some caution because one cannot infer from it that livestock components in broader based agriculture projects are performing less satisfactorily than other components. Since the rating system applies to the overall project we must assume (unless otherwise stated) that the performance of the livestock component is fairly represented by the overall rating. It is important to keep this in mind when one realizes that most of the lending for livestock is now made under component projects (Tables 1 and 2). The findings of the three studies which I referred to are still valid and describe the main problems which underlie poor project performance.

Overall performance is heavily biased downwards by poor performance, especially for Africa and to a lesser extent the East Asia and Pacific Region. It is important to note that the performance of Livestock projects in all other regions was similar to Agriculture projects in general. The bias is considerable because the volume of lending and the size of projects are relatively small in these regions.

In response to past deficiencies the lending program is now paying more attention to supporting smallholders, improving technical support services (i.e., extension, research, veterinary) institution building, cost recovery and privatization. Considerable emphasis is being placed on strengthening and restructuring veterinary services, on extension and research and encouraging the formation of Pastoral Association (for veterinary and resource management) in the Sub-Saharan Africa. Dairying is receiving renewed emphasis; dairy projects are under consideration, for example, in Sudan, Kenya, Madagascar, Uganda and Zambia. Furthermore, livestock have an important role in the "Bank's Areas of Special Emphasis", namely: poverty alleviation, food security, protecting the environment, private sector development/public sector reform and women in development. There are, for example 26 Sub-Saharan African countries with extension and 17 with research projects, although the primary focus in these projects is on agriculture and not livestock. Much more needs to be done to strengthen livestock extension and research.

In addition to new initiatives in project design more efficient economic policies are being emphasized in structural and sectoral adjustment lending. These include more realistic exchange rates, free market policies, privatization and cost recovery.

IMPORTANT FACTORS AND EXAMPLES

Profitability

It is important that long-term markets are available (and verified) to ensure profitability. Financial projections and rate of return calculations should be conducted with care. Although the World Bank places great emphasis on project analysis it is surprising how often projects fail because of inadequate profitability. Since livestock development is a long term activity, inflation and as a consequence high interest rates cause serious cash flow problems during the earlier project years. In addition to problems with prices and markets (which should be verified with care) technical and production coefficients are often over-optimistic. Unless judgements are made by experienced operators they are inclined to reflect what is technically feasible, or what is feasible in the developed world, without due allowance been made for differences in management capabilities, technical services or the availability, cost and quality of input supplies. For example, in African ranching or dairy projects it was assumed by all donors in earlier years that parastatals could achieve production coefficients similar to those obtained on good commercial (settler) farms. With hindsight we know this was not the case — parastatal coefficients were similar to those achieved in the traditional sector and in addition parastatals suffered from political interference, overstaffing and price regulation as well as poor management (including financial management).

Profitability is also determined by the level of capital investment. Smallholders have a substantial advantage in most cases over large commercial farms. Combined with low opportunity cost labour (family) this can give smallholders a distinct advantage over large farms and parastatals. Consequently their financial viability is considerably less threatened by low production coefficients, prices or management skills.

Economic Justifications

Even if an enterprise is financially attractive this may not mean that it is economically attractive as measured by the economic rate of return (ERR). The ERR omits subsidies and taxes. Border prices (free world market prices) are used to value inputs and outputs. Consequently, a satisfactory ERR is a much better indicator of long-term sustainability than the FRR. The ERR is particularly important in countries where free markets are not permitted to operate and where the exchange rate and prices are seriously distorted.

Government Economic Policies

An overvalued currency is without doubt the most serious constraint on orderly sustainable livestock development in many countries. It is particularly serious for countries with an export surplus or the potential to develop exports. There is for example, an excellent market in the Middle East for live animals (especially sheep and goats) but Somalia, Sudan and Ethiopia have not been able to exploit this market, except to a limited extent, because exchange rates have been grossly overvalued. Conversion of earnings at the official exchange rates can be equivalent to a tax of 100% – 300% on livestock exports.

When one considers the stimulus that a 100% real increase in livestock prices would have on producers willingness to use new techniques and increase production, one begins to appreciate the magnitude of the overvalued foreign exchange problem. Governments are reluctant to move to a more realistic exchange rate because this would affect local consumer prices, especially for city consumers who wield a disproportionate political influence.

If the exchange rate is grossly overvalued the conditions are automatically created for a flourishing smuggling trade. Large numbers of animals (especially sheep and goats) are presently smuggled live to the Middle East from East African countries and large numbers of cattle are smuggled (walked) into Kenya from adjacent countries. If these conditions are not corrected, projects, which are aimed at improving livestock and meat marketing are doomed from the outset. This is usually true even if the actual marketing operation is in the hands of private traders with the public sector role confined to the provision of marketing and processing facilities on a fee for services rendered basis.

Although, it can be argued that smuggling has positive economic features it is a costly and inefficient way to conduct business. Government is denied access to revenue and foreign exchange which could have been generated by official exports.

A grossly overvalued exchange rate has an additional negative effect in that the real costs of inputs that require foreign exchange are not reflected in the price paid by farmers and are therefore, not recovered even if a full cost recovery policy (local currency) is in place. This is an important issue and common difficulty with revolving funds for veterinary drugs and medicines in donor supported projects. The adoption of a realistic exchange rate appears to be the solution to this problem. The Bank is addressing the exchange rate issue through its structural adjustment program, albeit with mixed results. Sustainable livestock development will, in many countries, depend on the success of these efforts especially if success is directly related to the availability of foreign exchange as would, for example, be the case for a veterinary project importing drugs and medicines. There will always be a problem of this nature unless a country's currency is freely convertible at realistic exchange rates.

Government price controls are a major cause of failure for projects which rely on Government or parastatal production, processing and marketing; they can also affect the private sector if rigorously implemented. Overall effects can even transcend national boundaries. Livestock projects in East Africa were negatively affected, to a major extent, by government price control policies — even private sector ranches as well as parastatals. Livestock specialists and planners should not feel too guilty about parastatal failures because these policies had their origins in economic management theories and embraced all sectors (e.g., coal mining, car manufacture, and transport (in European countries) in those heady economic planning days.

Technology

Sustainable technology must take the realities of the country in question into consideration. Too often attempts are made to transfer technology from developed countries without realizing that they are unsustainable because labour/capital ratios are completely different and support services less developed; not only technical ones but also electrical, mechanical, manufacturing, communications and transport.

Investment in large commercial state or parastatal dairy production and processing in Africa and elsewhere is a good example of inappropriate technology. Smallholder dairying requires very little capital investment and, in addition, can usually utilize low opportunity cost labour. Consumers in many countries are not prepared to pay the extra costs of heat treating, packaging and distributing milk. Systems which are based on the sale of unprocessed milk (either fresh or sour), distributed from door to door by farmers or local vendors, are much more robust from an economic standpoint. Consequently dairy projects based on pasteurization and sale of milk in bottles or packages can only be justified if the milk shed is located at a considerable distance (at least 50 km) from the city market. Even where pasteurization is justified by transport distances considerable savings can be made by selling bulk milk to local vendors; as an alternative to a marketing system which may be doomed from the outset by heavy packaging and distribution costs.

The key mistake in attempting to transfer modern dairying practices to the developing world, arises from a lack of appreciation of relative costs of labour and size of incomes which by comparison with the developed world are usually 50 to 100 times smaller. The relative cost of labour must also be kept in mind when designing appropriate Artificial Insemination and other technical and input supply services. When one considers that the average workers daily wage is only equal to about 2-3 litres of milk, it is clear that every effort must be made to utilize cheap labour and economize on fixed capital (buildings, machinery and vehicles) as well as economizing on foreign exchange expenditures. These issues are discussed in more detail in a World Bank report entitled "Dairy Development in Sub-Saharan Africa" which will be published shortly in the World Bank's Technical Paper series.

The importance of grading-up, by crossbreeding, to achieve good dairy merit, must be emphasized. Crossbreeding is the cheapest and least risky approach. It avoids problems associated with acclimatization and minimizes the disease risk.

The record shows that mortality rates are unacceptably high for European breeds imported to tropical/sub-tropical regions and the number of animals imported should not exceed what is needed to establish an efficient crossbreeding program.

Processing technology, for milk and dairy products, should concentrate on simple manual system or ones with minimal mechanization — for milk separation, butter churning and small scale village cheese manufacture. Butter and cheese manufacturing is particularly important in areas which have a pronounced seasonal milk surplus and that are located far from liquid milk markets. Milk powder manufacturing plants can rarely, if ever, be justified because of their high capital and operating costs and the availability of subsidized milk powder on world markets. Large processing plants for meat and poultry are also difficult to justify. Small abattoirs that provide slaughtering facilities for a fee are the best solution if the number of animals involved can justify moving beyond the private butcher. Housewives would usually prefer to use their own labour to dress fowl rather than pay high processing charges. Furthermore, live chickens are much easier to store and can be killed when needed. It is interesting to note that this system is still operating in Taiwan where per capita incomes are several times higher than in most developing countries. In my experience we should look to Asia for appropriate models for processing and marketing livestock. Furthermore, there is now ample evidence to support the proposition that responsibility for marketing live animals and livestock products should be delegated to the private sector. If the public sector has any role it should be confined to the provision of infrastructure (markets, railway stockyards and lairage at ports) and, in addition, help to facilitate these activities by reducing red tape, improving telephone and telex services, abolishing taxes, and facilitating veterinary certifications for exports. The working capital that is needed in the livestock and meat trade is extremely large and Governments must ensure that adequate amounts are available through the banking system.

Feed Supplies

The sustainability of smallholder livestock depends, in large measure, on feed supplies. Virtually, all smallholders produce forage and crop by-products and therefore dairying and cattle/sheep fattening enterprises can be undertaken by virtually all farmers. Furthermore, even if the quantity of feed is small it usually has little if any cash value (except when sold or bartered to nomads). Since home produced feed costs little, if anything, smallholder livestock farming is unburdened of one major risk-that associated with large fluctuations in feed prices and reliance on purchased feeds.

Likewise smallholder pig and poultry systems flourish on farms producing grains (e.g., maize, rice) and crops and cereal by-products (e.g., rice bran). Robust smallholder pig production was a feature of Danish agriculture until very recently (a flourishing pig production industry based on home grown barley) and robust smallholder pig production still flourishes in Poland (based on home grown potatoes and cereals) and in Yugoslavia (based on home grown maize and wheat pollard from home grown wheat). Likewise smallholder pig and poultry production flourishes in Asia because all smallholders have access to cheap rice bran. Rice milling is usually dispersed through numerous small villages and farmers receive or buy back most of the rice bran from their own crop; rice bran stores badly because its oil content is high and it is therefore unattractive to feed compounders. It must be fed fresh and since rice milling is a continuous process rice bran is constantly available to smallholders.

These traditional smallholder livestock systems are virtually risk free and will survive (as they did in Western Europe) until wage levels are such that their contribution to family income becomes insignificant or when they are incapable of providing an economic labour wage. For example, in Asian rice systems, a man can manage a flock of about 150 ducks grazing on rice paddies. This system persists when the cost of feed saved, by scavenging, is sufficient to justify a man's wage but the practice has collapsed in countries where wages have surpassed this level. These examples are given in order to stress the point that one needs to be vigilant and fully understand the implications of the underlying economic realities as well as their influence on switching points in farming systems (e.g., from manual to machine milking).

Smallholders are economically more robust than large commercial producers because they are much less subject to feed and product prices. They should be encouraged to the fullest whenever possible and will persist until gradually surpassed by economic development. Feed is normally the binding constraint on smallholder production and projects designed to assist this sector should always incorporate a well thought-out feed and/or forage component.

Commercial sustainable pig and poultry industries can be developed on imported feeds in countries with a deficit in these products (e.g., Taiwan, the Philippines and Korea). This is possible because it is much cheaper to transport feed grains than meat and poultry products. However, if the industry is to survive in the deficit country production standards and efficiency must be comparable with those found in developed countries. Developing countries usually has a substantial advantage in labour and construction costs and, in addition, the fertilizer value of waste products is usually much higher and as a consequence the waste disposal problem is minimized.

Although somewhat surprising, it is worth noting that few countries have a well thought out strategy for meeting their short, medium and long term feed supplies (energy and protein feeds). One needs only to look to the continuing chronic feed protein deficit in Eastern European and Asian countries to realize the full magnitude of the problem. Perhaps FAO and the World Bank could play a more substantial role in rectifying this situation. The gains that can be achieved, if enlightened policies are put in place, are phenomenal and clearly evident from the expansion of pig and poultry production, on imported feeds in many developed and developing countries.

Sustainable Support Services

Sustainable smallholder development depends on sustainable technical and other services. Sustainability of services is largely a function of cost recovery. Even where the principles of cost recovery and privatization are accepted the task of developing sustainable institutions to deliver these services is a formidable one. Considerable investment will be needed in institution building, technical assistance, management training and staff training at all levels. Grass-root farmer organizations (i.e., pastoral association and cooperatives) need in most cases to be established and farmers trained to own, operate and manage them. This is a task that goes much beyond the life-span of the typical livestock project. One should look to Operation Flood in India and village agriculture/ livestock cooperatives in Taiwan to see what can be achieved in the mature stage, and to initiatives for restructuring veterinary services and establishing pastoral associations in Sub-Saharan Africa to see what can be achieved at the earlier development stages.

In recent years the Bank has placed much more emphasis on institution building and the provision of services (e.g., extension, research and veterinary), in full cognizance of the importance of these to sustainable development. Although this is not the place to discus these important services, it is pertinent to point out, that in most developing countries livestock services are either non-existent, weak, non-effective or absent. The design of efficient affordable services is a difficult task which calls for innovation, the rejection of old nostrums and careful cost/benefit analysis to establish affordability, a pre-requisite for sustainability. While donor agencies can assist this process, governments must ensure that coherent policies and strategies are implemented to avoid confusion and to save time and resources. Donors must ensure that their activities and projects are consistent with the policies and strategies which have been set by Government.

Animal Health Services are of paramount importance to the sustainability of smallholder system but especially to dairy farmers using disease susceptible crossbreds. Although good progress has been made on conceptualizing restructuring and cost recovery most livestock farmers in Sub-Saharan Africa have still to 'make do' with, at best, a rudimentary and, at worst, a totally ineffective service. In Ethiopia, for example, the expenditure on veterinary drugs and medicines is only about 5% of the amount veterinarians estimate would be economically justified on the basis of epidemiological studies. It will be extremely difficult to rectify this situation even though the principles of restructuring and privatization are fully accepted. A major effort is needed to train veterinary field assistants that will be employed by service cooperatives, the front line institutions. Systems of bookkeeping and cost recovery must be developed and demonstrated which require time and a massive training program. A mechanism is needed to enable payments in local currency to be converted to foreign exchange to replenish imported veterinary stocks. In addition procurement and distribution must be streamlined.

Although one must assume that these problems can eventually be resolved one should question if the contribution or the role of farmers in administering drugs and medicines is adequately taken into account in present Sub-Saharan Africa models where veterinary assistants, selected from and paid by the traditional village community or pastoral association, are responsible for administering drugs, medicines and vaccines to livestock. When one considers that probably 90% of the veterinary drugs and medicines are administered by farmers in developed countries (as represented by farmer's expenditure), albeit in most cases under the veterinarian's instructions one begins to realize the importance of training the African farmer to play a much greater role in the administration of veterinary products. It would take an enormous increase in manpower, travel time and cost to replace the farmer's legitimate function in, for example, 'drenching' for stomach worms and liver fluke. Study and analysis is needed to sharpen the focus on these matters - perhaps FAO could help by establishing badly needed guidelines.

Environmental Sustainability

Despite some common misconception livestock projects normally make a substantial beneficial contribution to sustainable agriculture. In arid areas there is now ample evidence that irreversible degradation is not taking place on a large scale as a consequence of over-grazing. Rangelands generally recover when droughts give way to a wetter cycle of annual precipitation. The public clamour and fear of irreversible degradation appears to follow a similar cyclic pattern. The real environmental problem in arid areas is caused by erosion brought about by increased and continuous cropping as well as bush cutting (for fuel) which is, in turn, a consequence of population pressure. Overgrazing does exist and can cause erosion on slopes, but these effects are minor compared to those caused by human population pressure. Furthermore, grasses, forage, legumes and legume trees, introduced to provide livestock feed, are important builders of soil fertility and, in addition, provide ground cover which prevents or lessens wind and water erosion in susceptible areas (e.g., Ethiopian highlands).

THE IMPACT OF LIVESTOCK DEVELOPMENT ON ENVIRONMENTAL CHANGE

Environmental change, over the next 50–100 years, due to the warming effect of the accumulation of gases in the atmosphere will clearly influence man and his well-being and will introduce change to many areas of the world.

Although there has been modelling of future climate change, the major observation is that these models are as yet producing limited useful information. Computers have the ability to consider a multitude of variables but the computer is limited by the knowledge base and magnitude of the unknowns. In fact, it can be quickly ascertained that modelling, at best, is defining what is not known in terms of what will effect change in temperature, rainfall and sea levels, let alone sea currents, wind, sunshine hours, soil moisture and the incidence of pests and diseases of man, animals and plants.

Some prediction (no matter how uncertain) are welcomed by the population at large (e.g. increased environmental temperatures in Europe) but the uncertainties indicate that for any benefits there will also be major disadvantages, both economic and social. For example, high temperature and increased evaporation rates and the lower resultant soil moisture content, may result in the death of large areas of planted and natural forest in Northern Europe; rain in previously dry areas may lead to large scale soil erosion and wide scale flooding of major agricultural lands in the fertile delta country will occur if sea levels increase by 0.3 meters by 2050.

The major conclusion is that disadvantages are likely to outweigh any advantages and the unknowns make it essential to put into action technologies to slow green-house gas emissions and stabilise atmospheric gases as soon as possible.

The long lag time between gas production, mixing in the atmosphere and therefore warming together with the reluctance of governments to put into practice legislation to limit, in particular, carbon dioxide production from fossil fuels, suggests that there will be a rise in world environmental temperatures of 0.5 to 1°C in the next 25–50 years.

A major point is that no one country can point to a low level of "environmentally dangerous gases" production as being a reason for non-compliance with a general lowering of emissions. Virtually all countries contribute a small proportion of the greenhouse gas and it is, therefore, necessary for all countries to take action. In other words it needs action world wide and "every little bit helps".

A distinction must be made between temperature rise due to gas accumulation in the atmosphere and depletion of the ozone layer through reaction with atmosphere contaminants. The two are linked but the depletion of the ozone layer (this layer protects the animals/plants from deleterious dose rates of UV irradiation) is largely a result of reaction of the ozone with atmosphere contaminants. Ozone depletion is not highly related to the subject of environmental change and gas accumulation in the atmosphere. However, increased ultra-violet radiation at the earths surface will have major detrimental effects on plant growth.

THE GREENHOUSE EFFECT

A Simple Description

The greenhouse effect, or increasing world temperature, is clearly ascribable to the major industrial countries as some 50% of the increased retention of energy by the atmosphere is a result of the accumulation of carbon dioxide from combustion of fossil fuel. Industrialised countries presently use 70% of the world's oil production and it has been much higher in the past.

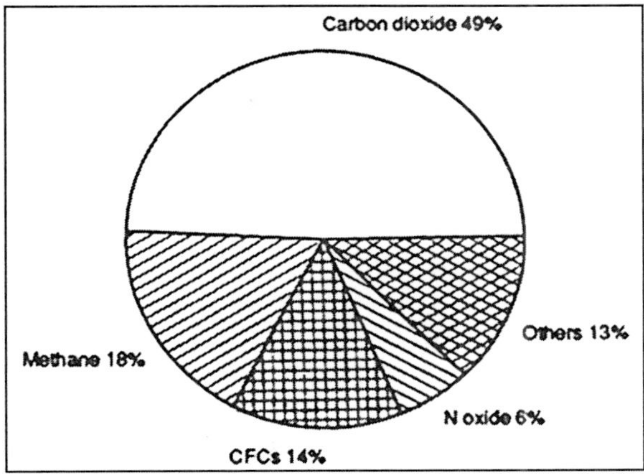

Fig. Relative contribution (%) of greenhouse gases to atmospheric warming.

The other gases that contribute to increasing temperatures arise from a variety of activities and some of the gases have been created by man. Methane is an important component of the increasing gases in the atmosphere and is the one most associated with animal agriculture. Methane has a thermogenic effect some 4–6 times that of carbon dioxide.

The rates of accumulation of methane and carbon dioxide in the world's atmosphere have changed dramatically in the last 10 years. Prior to this, the rise in world temperatures and composition of the atmosphere had changed little, but is now in what appears to be an exponential period. Undoubtedly the contamination of the atmosphere with carbon dioxide, methane and other gases must be stabilised or the future of the earth is threatened.

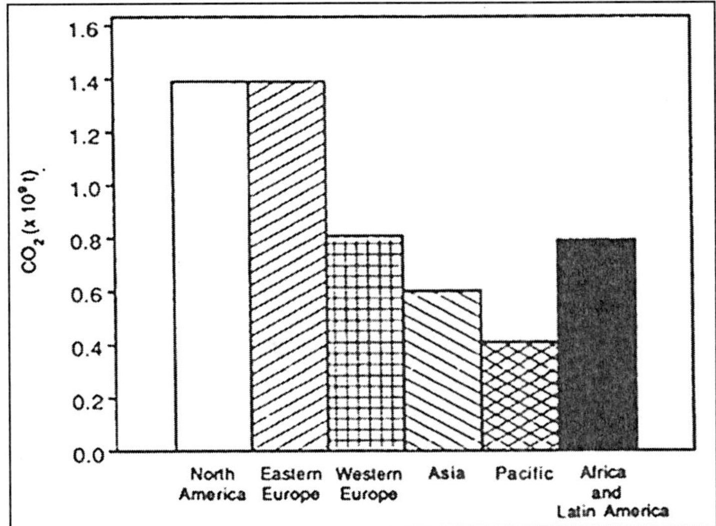

Fig. Relative contribution by continent to the emissions of carbon dioxide.

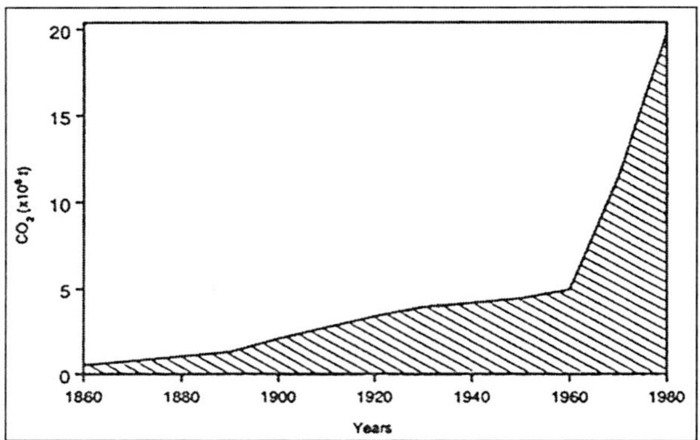

Fig. Trends in emissions of CO_2.

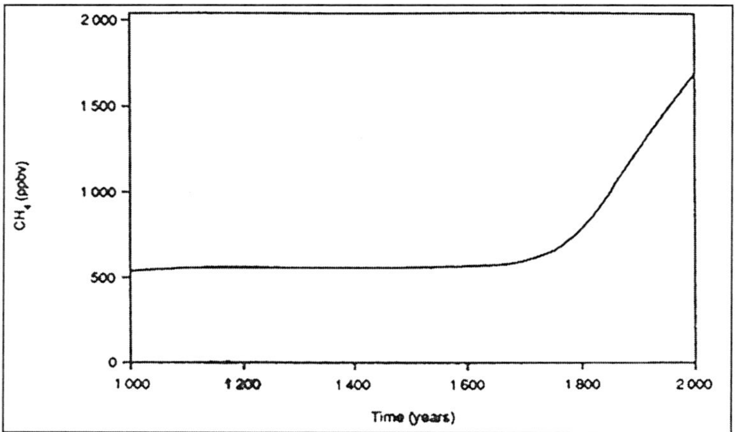

Fig. Trends in atmospheric methane accumulation.

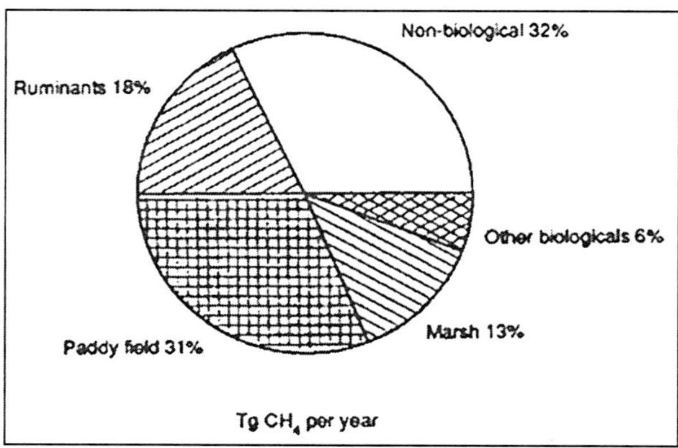

Fig. Relative contribution of biological resources to the global production of CH_4 in the atmosphere.

Methane production appears to be a major issue although it presently contributes only 18% of the overall warming. It is accumulating at a fast rate, and is apparently responsible for a small proportion of the depletion of the protective ozone layer. Methane arises largely from natural anaerobic ecosystems, rice paddies and fermentative digestion in ruminant animals.

Animal Agriculture and the Greenhouse Effect

Animal production plays four important roles in the release of gases into the atmosphere:

- Directly through production of methane in fermentative digestion of ruminants,
- Indirectly when a proportion of the faecal materials decompose anaerobically,
- Indirectly through CO_2 production from fossil fuels to provide the production and marketing infrastructure and inputs such as motorised transport, fertilizers, herbicides and insecticides,
- Through the clearing of forests and range lands, the timber on which was a natural sink for carbon dioxide.

Ruminants in 'natural' production systems are inefficient and, in general, production increases depend on an expansion of numbers. There is a growing appreciation that efficiency per animal can be improved many fold with simple technology inputs which would have an impact on all four aspects of the contributions to global warming.

The most important approach to be discussed in relation to the amelioration of greenhouse gas production by ruminants is to increase the efficiency of animal production from available resources and develop the capacity to produce more from less animals.

REQUIREMENTS FOR ANIMAL PRODUCTS

Developed Countries

With largely stabilised human populations in the industrialised countries and a high standard of living generally, the demand for animal products has plateaued or declined and the emphasis in animal agriculture is on the production of higher quality products. This has led to a dependence of ruminant systems based on concentrate feeds, a higher production rate per animal and reduced animal numbers. For example, there has been a marked increase in milk production through "better nutrition" relying on concentrate diets and genetic improvement of dairy cows. The result has been a marked reduction in the total number of animals required to supply local milk requirements. These increases in production per animal in North America and EEC countries that have arisen from simple technology.

The demand for food in the industrialised world is stable but production is increasing at 1.5% per annum. Policies to take land out of farming or to limit farm management options and animal numbers has reduced overall animal numbers, but meat/milk production has been maintained by technologies that increase the efficiency of production. From a global warming view point this is mostly advantageous, although there is a very high environmental cost of feeding high grain based diets to ruminants. Such concentrates can only be produced by high fossil fuel inputs and is often only economic because of subsidies. Les intensive production systems depending on grass or grass products in developed countries exhibit some of the same problems of low efficiency as those animals in developing countries which are restricted to roughage based diets:

The Developing Countries

Throughout the last 30 years crop and livestock production has more than doubled throughout the third world, although there are large differences between regions, however, increases in human population, urbanization and improved income levels have increased demand for food to such an extent that surpluses are still rare. Imports of meat, milk and cereal grains in most developing countries are increasing by about 10% per annum. The demand for animal products will continue to increase for some time, if the patterns of food consumption by people in developing countries follows the patterns that occurred in developed countries.

Increases in animal production in the developing countries has been mainly a result of increasing animal numbers. The lack of increase in efficiency of animal production is well documented and is emphasised by the average milk yields per cow over the 10 years from 1976–1986. This low productivity is exacerbated by long calving intervals and a late age at puberty in the cows in developing countries. It should be emphasised however, that it is also a feature of ruminants fed low quality forages in any country.

New feeding strategies for animals fed on low quality forages (e.g. crop residues, tropical pastures etc.) coupled with better genotypes, improved management and disease control, particularly in India (NDDB, 1989), has changed this situation enormously.

Table. The change in the average milk yield per cow in industrialised and third world countries.

Country/ Region	Average Yield/cow p (kg/year)		Percentage Increase (%) p (1976 to 1986)
	1976	1986	
North America	3,250	5,200	60
EEC Countries	2,900	4,100	35
Asia	620	700	13
Africa	322	354	7

A major difference in approach to feeding ruminants in tropical developing countries that has been recently developed has the potential to revolutionise ruminant production from forages. For example enormous increases in milk production can be achieved in the tropics without the use of 'fossil-fuel-expensive' grain based concentrates, relying rather on byproducts of agriculture.

These are the only truly available feed resource for the large ruminant populations in the foreseeable future. The low inputs of concentrates into such systems and levels of production that rival those achieved in the industrialised world, provide major indications for change in management of both developed and developing countries alike.

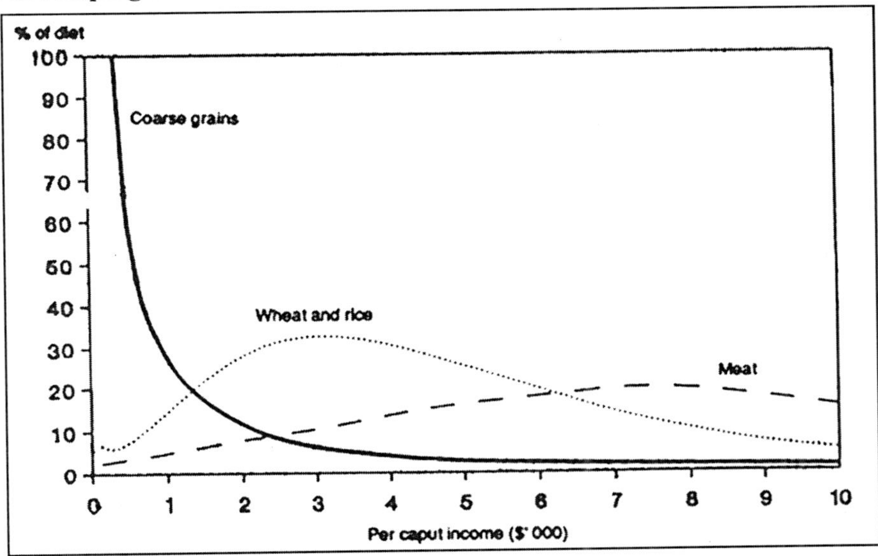

Fig. Food consumption/percentage of diet for meat, wheat & rice and coarse grains.

Population Densities of Large Ruminants

The ruminant populations are not likely to increase in the industrialised world and, with ever increasing technology inputs, numbers are set to decline (as has already occurred with milking herds) - but continuing technology inputs will be highly fossil fuel dependent.

Conversely, there is likely to be a huge increase in demand for food products in the countries that are developing, of which a greater proportion of the demand is likely to be for animal products. To meet this demand local production must be increased to an extraordinary extent and it is most desirable that the impact on the environmental contamination is minimised. of essential nutrients for microorganisms. The ranges of Y_{ATP} are shown for: The demand for draught power in countries with large numbers of small farmers is likely to expand rather than contract in the future and must not be left out of any considerations.

This large group (80 million draught oxen in India alone) receives the least inputs and yet they are probably one of the major factors in food production and they limit the use of costly fossil fuels in developing countries. Environment-friendly development of livestock production systems demand that the increased production must be met by increased efficiency and production per ruminant and not through increased numbers. The need to increase numbers would put huge pressures on many resources including forests and land that might be afforested.

METHANE PRODUCTION

Methane Production from Ruminants

Global Methane Production

Methane gas accumulation is at a rate of 1% per annum and methane contributes 18% to global warming.

Ruminants from all species produce a relatively small proportion of the global production i.e. 15–20% of the total methane generated. However, the domestic ruminants represent one of the few sources that could be manipulated. It is estimated that beef and draught animals contribute 50%, dairy cows 19% and only 9% is from sheep. The methane is generated largely in the fermentative digestion of feed by microbes in the rumen.

An approximate even split of numbers of ruminant animals between the developing and developed countries; that large ruminants contribute the greatest to world methane accumulation, and that other domestic and wild herbivores and man contribute an insignificant proportion.

The simple approach to the calculations adopted by Crutzen *et al.* (1986) may, however, be a little misleading although they are the best estimates available.

It is most important to emphasise that it is the rates of methane production per unit of product produced over a life time which is important in identifying where it is possible to make a major reductions in methane emissions. This is the major philosophy developed in the rest of this presentation.

Productivity of ruminants fed poor quality forages

The vast majority of ruminants in developing countries and a major proportion of the national herds of industrialised countries are supported on the by-products of agriculture or graze forages of relatively poor nutritional value. In general, growth rates, milk production and reproductive rates in these systems depending on forage of variable quality are extremely low compared with the genetic potential of these animals (mostly about 10% and rarely exceeding 30%).

Mostly cattle grow to maturity or slaughter weight over 4–5 years, cows produce their first calf at 4–5 years and then, on average, every two years.

Milk production on these feeding systems is often below 1000 litres/lactation. Cows may be kept largely to produce draught oxen and in some specialised systems they are kept for the production of dung (which is valued as a fuel) and a number of other minor purposes (e.g. as a investment, for recreation and for religious purposes).

Table. Estimates of methane emissions from animals.

Animal Type and Region p(X10^{-6})	**World Pop.**	**CH_4Prod.p (kg/hd/yr)**	**Total CH_4 pProd.**
Cattle:			
Developed countries	573	55	31.8
Developing countries*	653	35	22.8
Buffaloes	142	50	6.2
Sheep:			
Developed countries	400	8	3.2
Developing + Australia	738	5	2.4
Goats	476	5	2.4
Camels	17	58	1.0
Pigs:			
Developed countries	329	1.5	0.5
Developing countries	445	1.0	0.4
Horses	64	18	1.2
Mules & Asses	54	10	0.5
Humans	4670	0.05	0.3
Wild ruminants and			
large ruminants	100–500	1–50	2–6
Total			76–80

Note:

* includes Brazil and Argentinap** total estimate for emissions from domestic animals has an uncertainty factor of ± 15%

Slow growth, low milk yield and poor reproductive performance results in poor feed conversion and a large methane output relative to product output.

Methane production in ruminants fed 'poor quality' forages

Methane output relative to product output of ruminants depends on two factors:

- The efficiency of fermentative degestion in the rumen, and
- The efficiency of conversion of feed to product (e.g. milk, beef, draught power).

Efficiency of rumen fermentation

Digestion of feed by ruminants depends on a diverse group of micro-organisms in the rumen. These organisms ferment feed materials into volatile fatty acids (VFA) a process that produces methane (CH_4), carbon dioxide (CO_2) and utilises the energy (ATP) derived to convert feed to microbial cells. The partitioning of the feed components into VFA or microbial cells and the release of CH_4 and CO_2 dependspon a number of factors. Feeds that allow a high efficiency of microbial cell synthesis produce low amounts of methane per unit of feed digested.

With cattle fed on a poor quality forage a number of essential microbial nutrients are usually deficient in the diet and microbial growth efficiency in the rumen is low. In these conditions CH_4produced may represent 15–18% of the digestible energy and correction of these deficiencies may reduce this to as low as 7%. The relationships between products of fermentative digestion and the efficiency of the microbial ecosystem.

Efficiency of feed utilisation by ruminants fed crop residues or other fibrous feeds

Liveweight gain. Research in the past 20 years has clearly illustrated that supplementation of cattle on low quality forage based diets effects productivity through increasing efficiency of feed utilisation.

A mixture of nutrients as can be supplied for instance in a molasses urea multi-nutrient block/lick ensures an efficient microbial digestion in the rumen. Also small amounts of protein meal that are directly available to the animal (i.e. bypass protein) stimulate both productivity and efficiency of feed utilisation.

Traditional feeding standards are based on the metabolisable energy (ME) content of a feed. The general relationship between ME/kg of feed and growth (g gain/unit of ME intake). The results of a number of feeding trials with cattle on straw or low quality pasture and silage based diets supplemented with protein meals.

The efficiency of growth and methane production. These data clearly show the massive reduction in methane production per unit of liveweight gain that is possible by using a strategic supplementary feeding system that accommodates the requirements of the rumen organisms and balances the absorbed nutrients to the animals requirements.

The effects on methane production of balancing the rumen and for protein supplementation calculated for experimental data with growing animals of Saadullah (1984). This indicates the massive potential reduction in methane production per unit of liveweight gain that can result from supplementation.

Provision of molasses urea blocks to draught oxen would have a major effect on methane production, reducing it to half the present production rate.

Milk Production

The same principles of supplementary feeding have been found to stimulate milk production of dairy animals. Without going into detail, the methane produced per unit of milk produced under traditional feeding of local dairy cows or imported Friesians in India or Friesians under temperate country management are production in relation to lifetime milk production and includes the effects of supplementation on age at first calving, intercalving interval and improved milk yield of supplemented animals.

Figure: Relationship between the production of microbial cells and volatile fatty acids and methane in fermentative digestion in ruminants.

The relative efficiency of the system (indicated as Y_{ATP}) is governed largely by the availability of essential nutrients for microorganisms. The ranges of Y_{ATP} are shown for:

A. - a relatively inefficient rumen (i.e. ammonia deficient),

B. - a 'normal' rumen with no deficient nutrient for microbial growth,

C. - a rumen free of protozoa with no deficient nutrient for microbial growth,

D. - the theoretical optimum microbial growth efficiency.

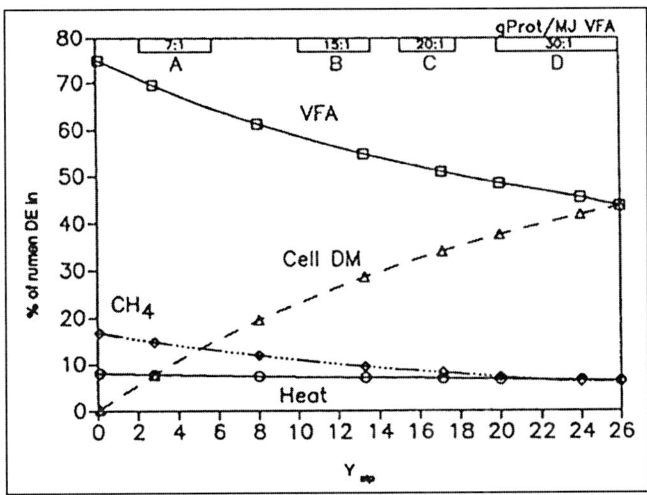

Fig. Schematic relationship between diet quality (metabolisable energy/kg dry matter) and food conversion efficiency (g liveweight gain/MJ ME).

The relationships found in practice with cattle fed on straw or ammoniated straw with increasing level of supplementation. Australia (◊, O, •), Thailand (Δ) and Bangladesh (□). Recent relationships developed for cattle fed silages supplemented with fish proteins (⊕) and tropical pastures supplemented with cottonseed meal (Godoy and Chicco, 1990) (*) are also shown. This illustrates the marked differences that result when supplements high in protein are given to cattle on diets of low ME/kg DM.

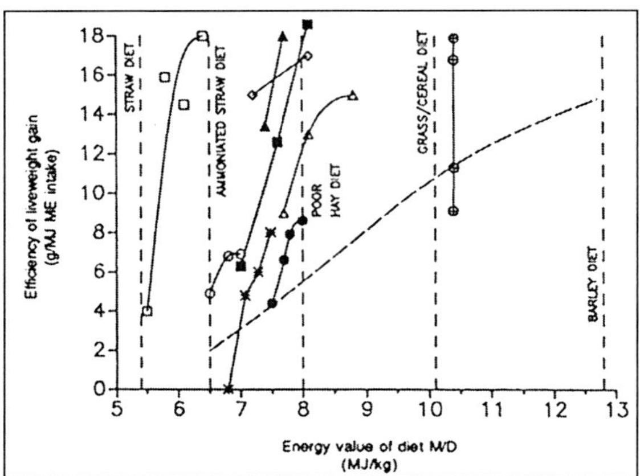

Fig. The relationship between the metabolisable energy content of a feed (M/D, MJ/kg) and the methane produced/kg gain.

The relationship shown by a broken line is based on the metabolisable energy system in practice in the U.K..

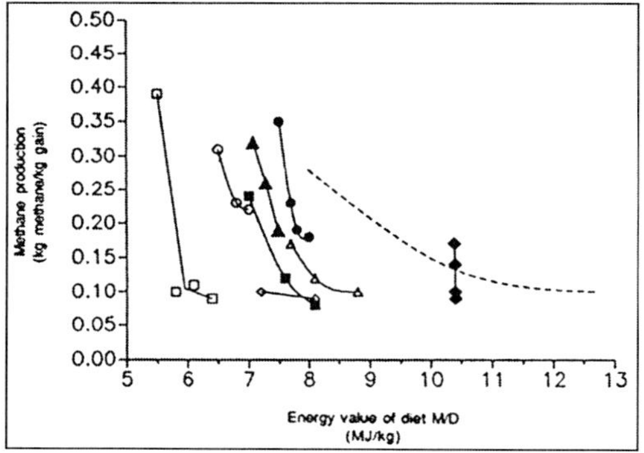

Fig. (A) The effects of improving the efficiency of rumen fermentative activity on methane production/kg of digestible energy consumed. (B) The production of methane/kg gain in supplemented cattle (feed conversion efficiency (FCR) 9:1) or unsupplemented cattle (FCR=40:1) fed straw based diets.

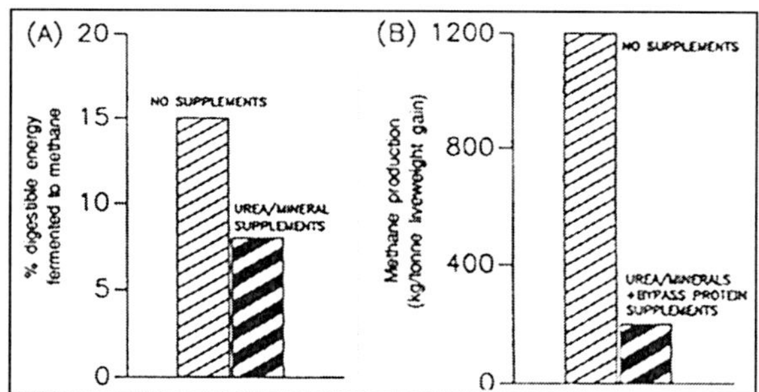

Fig. The methane produced per unit of milk production in unsupplemented (fed traditionally) or supplemented (new feeding systems) cows in India with moderate levels of production.

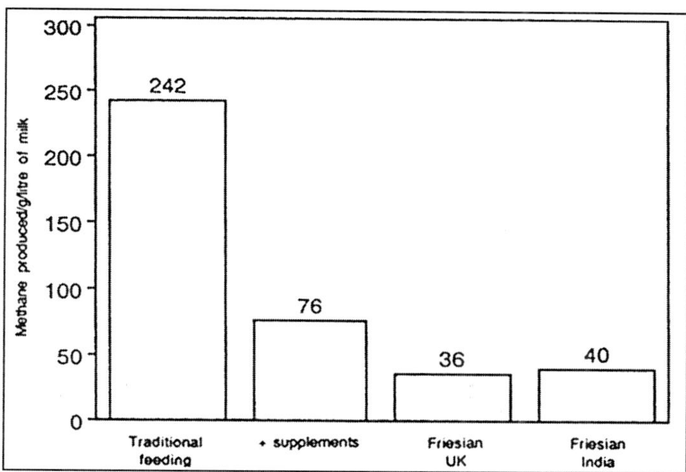

Methane gas content of the atmosphere is increasing at 1% per annum. To stabilise methane in the atmosphere, global methane production needs to be reduced by 10–20%.

Large ruminants produce some 15–20% of the global production of methane. Ruminants on low quality feeds possibly produce over 75% of the methane from the world's population of ruminants. Supplementation to improve digestive efficiency in these animals could at times halve this methane production per unit of feed consumed. Together with supplementation to improve efficiency of feed utilisation and increase product output may thus reduce methane production per unit of milk or meat by a factor of 4–6.

Provided animal numbers in national herds are decreased as demand is met, the production of methane from the large populations of animals fed poor quality forages could be reduced to below 50% and perhaps even to as low as 20% of its present rate.

It is probably not feasible to restructure the cattle industries of the countries involved. The socio-economic implications are difficult to predict, but the start made in India with up to 1 million head of cattle, and with a potential to reach 7–10 million animals in the immediate future, is highly desirable because it increases milk production with a concomitant decrease in actual methane production and methane production per litre of milk.

Social, anthropological, economic and political considerations have been the major determinants of 'aid programmes' but the growing 'environmental crisis' due to increasing content of gases in the atmosphere leading to global warming is likely to dominate many issues confronting aid-agencies in the future.

The achievements in India can be repeated in most countries that depend on poor quality forages and even silages for ruminant production. India is fortunate in having large amounts of protein meals as by-products which can be used for animal production. In other countries protein meals may be scarce or under-utilised. The future for increased animal production in these countries is to identify protein sources and proceed to find economic mechanisms for producing meals that contain a high proportion of bypass protein. The developed world may need to encourage this by subsidising such activities or by refraining from draining _these protein sources from the developing world to support subsidised over-production in the industrialised countries. The world trade in protein meals.

Table. The trade in oilseed cakes between industrialised and third world countries.

	Imports	Exports	Net Imports	Net Exports
Industrialised countries	18.7	9.4	9.3	-
Developing countries	1.7	10.8	-	9.1

The increases in efficiency of animal production that result from the new nutritional strategies will, if widely applied, increase animal productivity to the extent probably needed by the expanding human population in developing countries. However, it needs a retraining of technologists in these countries to absorb the new concepts and a move away from temperate country teaching of ruminant nutrition.

The tropical climate can be used to the advantage of the developing countries where ruminant production can be more efficient.

Acceptance of these feeding strategies could reduce the need for land clearing and pasture establishment in the fragile areas of the world which have been so prone to erosion following clearing. It may also allow for considerable change in the use of such pastures with reforestation as a highly desirable first option. In addition, reduced world ruminant populations would reduce incidental release of methane from decomposing dung and reduce the inputs of fossil fuels required in much of the infrastructure.

2

Sustainable Animal Agriculture and Future Challenges

If a sharply rising population in the developing countries is to achieve higher real incomes and a better quality of life, agricultural output must rise more rapidly than population growth. Further, if this rise in agricultural output is to be sustained over time, the natural resources which provide the basis for such output must be preserved and new technologies offering higher productivity must be developed. Increasing livestock production is an important component of this process, both because developing country consumers are expected to spend an increasing share of their rising incomes on livestock products, and because taking advantage of favourable livestock-crop production interactions is one approach to a more efficient, sustainable agriculture.

Despite the generally favourable effect which the use of livestock has on agricultural resources, livestock are also a component of several unsustainable production systems. Among the most widely cited include:

- The overgrazing of arid and semi-arid lands, leading to range degradation and productivity decline,
- The destruction of humid rainforests for the establishment of pastures which degrade soils and quickly cease production,
- The pollution of watercourses from animal wastes resulting from intensive dairy and meat production, leading to reduced production and consumer welfare elsewhere,
- The production of methane gas as a result of ruminant livestock production contributing to the threat of a global "greenhouse" effect, and
- And the pollution of soils, watercourses and subterranean water supplies from the application of fertilizers, herbicides, and pesticides in the production of livestock feed grains.

Increased attention to the issue of sustainability in livestock production is important if these production systems are to be improved. This chapter is concerned with establishing a number of guidelines for optimizing livestock production in developing or less developed countries (LDCs), consistent with achieving sustainable agricultural production. The paper focuses on economic factors (the author's specialization) though it contains some thoughts on the role of political and social factors. The paper first provides an overview of recent achievements concerning livestock production in developing countries livestock production. The paper then briefly discusses how different types of livestock are used as capital goods to produce multiple livestock products, and how the total amount of livestock capital and its use is highly responsive to the economic incentives provided to producers. The paper then analyzes several of the more prominent types of government interventions in livestock markets, discussing their probable effects on economic welfare. An emphasis is placed on the need for research. Finally, several problems involving sustainability in livestock systems are discussed.

RECENT EXPERIENCE WITH LIVESTOCK IN THE DEVELOPING COUNTRIES.

Within numerous national and international agencies, including FAO and the World Bank, concern has been expressed with the progress, or lack thereof, recently observed in livestock development. What does the record show? Have efforts to encourage livestock development been successful and what are the prospects for the near future? What policy changes could improve the situation?

Consider first the evidence regarding the rate of growth of livestock output in LDCs (Less Developed Countries), both absolutely and relative to population growth. One comprehensive study notes that food produced from livestock grew more rapidly than population for most products in Asia, North Africa/Middle East, Sub-Saharan Africa, and Latin America between 1961-65 to 1973-77. These results, with the exception for eggs, held for meat in all regions apart from Sub-Saharan Africa, but for milk only in Latin America. On average, the output growth rate for meat, milk, and eggs in developing regions were 2.9 percent, 2.5 percent, and 5.3 percent, respectively. Thus, per capita meat output grew about 0.5 percent per year during this period, though there were enormous variations across countries. Per capita milk output remained stagnant whilst per capita egg production rose rapidly. This record must be considered mediocre, especially since the consumption of livestock products generally rose at a more rapid rate, leading to rising imports or falling exports.

A more recent study confirmed this result for milk (Jarvis and Saha, forthcoming). Milk output in developing countries grew at a steadily declining rate from 1960 to 1988, falling from 2.7 to 2.1 percent per year. However, the rate of output growth varied significantly across geographical regions and from one sub-period to another which suggests that milk production has been responsive to changes in the incentives faced by farmers. For example, although the growth rate of milk output was low and declining in Sub-Saharan Africa and

Latin America (±2 percent) during the last decade where per capita incomes were declining; it was high and increasing in East Asia (nearly 3.5 percent in the last decade). The regional growth in milk output is highly correlated with regional economic growth, suggesting that demand for milk has been an important determinant of milk production. Indeed, growth was highest where the resource base (natural pastures) were limited and vice- versa.

Note that milk output in the developed countries, which still accounts for 75% of world milk production, also declined during the last three decades. Nonetheless, subsidized exports from developed countries contributed to rising LDC milk imports and, in general, discouraged milk output.

Prices

Comparable data on real livestock prices are relatively difficult to obtain for a large number of countries, but it appears that, among meats in LDCs, beef prices have been relatively constant, again with wide variation across countries. Domestic prices have often been tied, at least loosely, to international beef prices which determine the cost of beef imports or the value of beef exports. International beef prices have remained essentially constant during the last three decades. Within Latin America, for example, beef prices have risen in a number of major producing countries as consumption as risen more rapidly than production, forcing countries from a position as net exporters to self-sufficiency. These governments have gradually reduced interventions which depressed beef prices.

In contrast to beef, poultry prices have fallen steadily throughout the world, mainly in response to increased production efficiency resulting from: improved genetics, balanced feeds, veterinary drugs, economies of scale and better management. Poultry consumption has risen rapidly in response to its declining real price and is becoming the meat of choice in many LDCs, particularly in urban areas.

Milk prices have fluctuated widely in response to the variation in the level of export subsidies from developed countries, especially the European Community (EC). However, many LDCs have imposed tariffs or quotas on powdered milk imports to protect domestic producers. India, South Korea, Brazil, Chile and Colombia are examples and milk production has responded positively. However, as will be argued below, in each of these cases, technical change and increased producer efficiency has contributed more to the growth in output than have higher prices *per se*. Indeed, milk output in each of these countries has been growing more rapidly than domestic consumption and, as each country has approached self sufficiency, governments have been reducing domestic prices. Future growth in the milk industry in these countries will depend on the growth of domestic milk demand and/or each country's ability to export milk in an increasingly competitive world market.

Return on Investment

What has been the profitability of livestock investments? In a World Bank ex-post evaluation of 104 agricultural development projects which were wholly or partially devoted to livestock, roughly 60% of the economic rates of return (EER) exceeded 10%, while 34% were below 5% (The World Bank, 1985). The average return on livestock investments, weighted by the amount invested per project, exceeded 12%. Although these are only moderate returns relative to those estimated for other Bank investments, livestock projects have offered an acceptable return. More importantly, most of the projects which failed were in Africa, whose economic, technological and political context offered unusual challenges. More than two thirds of the projects in other regions performed satisfactorily.

Analysis of these projects suggested that the variation in their performance was determined by many factors. Factors which had particularly strong effect included the availability or lack of:

- Technological packages adequately adapted to existing farming systems,
- An economic context providing attractive producer incentives,
- An institutional capability for implementing the proposed project,
- Qualified technical personnel to design and implement livestock development programs,
- A government commitment to livestock development, meaning a genuine willingness to establish a stable policy framework and commit the required public resources,
- Political and economic stability, and
- The existence of clear property rights, which provide incentives for development and use of land and complementary resources.

Rather than suggest that there is something inherently difficult about livestock production, the World Bank's experience suggests that production depends strongly on the incentives faced by producers. These incentives, although strongly influenced by domestic income, population growth and by the international context within which countries trade, can be significantly affected by government policy. Technical change is an essential component of any long term programme to increase livestock output, yet investment in livestock related research in developing countries has been low and the lack of appropriate technology has been an impediment the growth to livestock output.

In summary, livestock output has been growing, but ruminant production (beef and milk) have been rising more slowly than monogastric poultry and pork. The trend in beef and milk prices has been relatively constant, whilst of poultry and pork prices have steadily declined. These figures suggest that technological change has been more rapid in the production of poultry and pork, allowing producers a favourable return while offering consumers a cheaper product. A higher level of technological change combined with protection of the resource base, is needed if ruminant output is to increase and provide consumers with more, cheaper and better products.

THE USES OF LIVESTOCK AS AGRICULTURAL CAPITAL.

Livestock contribute in many ways to national welfare in developing countries. On average, livestock account for half of agricultural output when bothe their direct and indirect contributions are considered. Directly, livestock provide food and non-food products (hides and skins) amounting to about 20 percent of agricultural GDP. Indirectly, they contribute another 30 percent by supplying essential inputs to agricultural production. Livestock convert crop residues, agricultural by-products and pastures on marginal lands (resources with limited alternative use) into a range of higher value products for subsistence and sale. The integration of livestock into cropping, via draught power and manure, increases the area cultivated, improves the timeliness of agricultural operations and helps maintain soil structure and fertility. As economic development proceeds, the use of ruminant livestock by small farms will shift away from the current emphasis on traction towards beef and/or milk production.

Because animal food products command high prices, they are usually consumed in greater amount by individuals having higher incomes, although, in some countries in Africa and Latin America they may be account for a high proportion of expenditure even in poorer households. However, even where the poor consume few livestock products, it is economically attractive for them to produce such products and exchange these for other foods which provide cheaper sources of energy and protein.

Both species and breeds vary in their capacity to produce different types of outputs and to utilize different types of inputs. Ruminants have the capacity to utilize low quality, bulky feeds such as pasture, crop and industrial by-products which have few alternative uses. Large ruminants (buffalo and cattle) also provide draught power and are by far the main source of milk. Small ruminants (sheep and goats) are generally more prolific, produce wool and hair in addition to meat and milk and can prosper under poorer range conditions than large ruminants. Small ruminants are also a more convenient household source of meat, barter and cash. They are also more easily stolen. In many situations, large and small ruminants are complementary since they utilize different forage species. Diversification of disease risks is further reason for running them jointly.

Monogastrics (pigs and poultry) are utilized primarily for meat production, although hides, feathers, down and manure are also important products. The principal economic advantage of non-ruminants is their ability to convert high energy/protein feeds into meat at a more favourable ratio than ruminants. Such feeds are often expensive if they are demand in for direct human consumption.

However, an increasing amount of high energy/protein feeds from crop by-products are now fed to non- ruminants and identification of cheap feed sources for monogastrics is an important aspect of livestock production in most developing countries.

In most countries there is a choice between two fundamental livestock production strategies a) the feeding of inexpensive, low-quality pasture resources to ruminants to produce meat, milk, wool, manure and draught power, and b) the feeding of high energy-high protein grains to non-ruminants for egg and meat production whenever the demand exceeds the amount which can be produced by ruminants from low-cost feed resources available. It is generally uneconomic to produce beef using feed grains, except where beef prices are unusually high, although the price of milk may justify such feeding.

Where pasture, forage or low-quality crop by-products are available (or can be economically increased), ruminants provide meat and milk at low cost. There is potential to increase pasture production in Latin America and so increase beef and milk production. A similar potential probably exists in much of Africa if trypanosomiasis can be controlled. In most other regions, pastures are limited and increased meat and milk production will have to come mainly from swine and poultry utilizing grains and high-quality agro-industrial by-products. Such production will commonly take place in large, industrial-type enterprises close to urban centres.

Despite the rapid growth in demand for poultry, smallholders in developing countries can be expected to rely on ruminants as their primary livestock assets because they use more efficiently the locally available, low quality feeds, and that they provide a wider range of products, particularly draught and manure, crucial to their overall farming system. In meat production, smallholders can compete effectively with larger commercial enterprises only to the extent that they have access to low cost farm resources, especially feed and labour, which cannot economically be sold off-site. Such low cost resources usually result from the integration of agricultural and livestock activities. Many smallholders will find it profitable to maintain a small number of other species to utilize that available feed which ruminants do not utilize efficiently, to provide diversity to the family diet and to provide assets which can be liquidated in smaller amounts.

THE ROLE OF GOVERNMENT IN DETERMINING LIVESTOCK PRODUCTION INCENTIVES.

Livestock are capital goods which are highly mobile and can be liquidated rapidly if economic incentives are unattractive. Livestock production can thus be strongly affected by government policy, both insofar as it affects: a) the prices which farmers face for their products and for the inputs they purchase, b) as it affects property rights (especially rights to land ownership and use), c) the development of new technologies, d) agricultural extension, e) the availability and terms of credit, f) animal health and sanitation, and g) infrastructure (e.g., roads, communications, and police and judicial services). Although it is impossible in a brief paper to cover the workings of all policies in all countries, a number of the most important policies affecting beef and milk in developing regions during the last two decades will be briefly analyzed.

As a general rule, economists believe that governments should work to ensure that the price paid to domestic producers of, say, beef or milk is equal to the international value of these products, i.e., the FOB value of exports and the CIF value of imports in a local port, plus or minus domestic transport costs to the site of production or consumption, respectively. The same rule is valid for establishing the cost of productive inputs. When this rule is not followed, national economic welfare is usually diminished, though there are some important exceptions.

Historically, many governments (e.g. Latin America) have tried to reduce the retail prices of beef in order to benefit consumers. However, lower prices to consumers usually require lower prices to producers, which lead eventually to lower output. Thus, the effort to assist consumers via lower prices is usually successful only in the short run, and may actually harm consumers in the longer run as output declines. Such policies have been adopted at least once by nearly every Latin American country during the last two decades. Although initially all consumers might have benefitted through the availability of beef at a lower price, this benefit would soon be lost for many as consumption is reduced or higher black market prices. Of course, producers lose in both the short and the longer term due to the lower price received. Note that the analysis is similar if prices are imposed at the producer or wholesale level.

A similar effect will be achieved if a country which exports beef imposes a beef export tax, reducing the domestic price of beef below the international level. The relationship between import and export on meat supply and demand where the import price of beef is higher than the export price. Given that the intersection of domestic supply and demand occurs at a price lower than the export price of beef (P_e), it will be profitable for the country to export beef. If the government allows free trade to occur, the domestic producer and consumer price will be equal to P_e, domestic consumption and production will equal C_o and Q_o, respectively, and exports will equal X_o.

Suppose now that the government imposes a beef export tax equal to t, where t is some fraction of the export value. The domestic price will decline to $P_e(1\text{-}t)$, causing an increase in domestic consumption, a decline in domestic production, and a decline in exports. Moreover, it can be shown that economic welfare falls by an amount equal to the two shaded triangles. These losses occur because consumers are now artificially induced to spend more on beef and because producers are discouraged from producing beef, thereby, diverting their resources into other activities which the country produces less efficiently. In addition, the reduction in the domestic beef price causes a significant redistribution of domestic income. Beef consumers gain because the price of the product they purchase has declined; the aggregate gain for consumers is approximately indicated by the area abdc. In turn, producers lose because the price they receive for beef has also declined; their aggregate loss is approximately area aefc. The government earns tax revenue equal to the area tX_1. Because of the shift in income distribution, consumers may favour an export tax even though total national welfare is diminished by its effect.

It may be noted that some countries which have imposed a beef export tax have also neglected to develop new beef production technologies, perhaps in the belief that the beef export surplus will remain. Over time, however, beef consumption in such countries may grow more rapidly than production (shown by a greater rightward shift in the demand curve than the supply curve). If so, exports will decline. Once exports have been eliminated, the country will be simply self-sufficient and further increases in domestic demand will lead to a rising domestic price until the import price, P_m is reached. Subsequently, beef imports will begin.

Milk, like beef, is politically a highly sensitive product and has also frequently been subject to intervention. Such intervention has sought two distinctly different ends. In many countries, governments have concentrated on restraining prices. The effect has often been to depress output and decrease milk quality.

For example, in Colombia the government has imposed price controls at various stages of the production and marketing chain for pasteurized milk, but not for crude milk and processed products. The expressed intent was to reduce the cost of pasteurized milk, thereby benefitting poor consumers. The results have been somewhat different. Price controls on pasteurized milk have led to a) higher adulteration and lower milk quality, b) a higher percentage of milk marketed as raw milk with attendant health risks, and c) a lower price and higher consumption of processed milk products (cheese, yogurt, ice cream, etc.) which are consumed mostly by upper income groups, at the expense of a reduced supply of pasteurized milk, which is consumed more evenly by all income strata.

A different situation is one in which the government has imposed an import tariff on milk to protect domestic producers. This policy was commonly used by developing countries during the mid 1980s when international milk prices were strongly depressed by subsidized exports from developed countries. A number of countries which had rising milk imports (eg. South Korea, India and Indonesia), made significant efforts to develop their milk industry to satisfy a growing national demand.

The effects of milk import tariffs. Where the intersection of domestic supply and demand lies above the import price of milk. If free trade in milk (powder) is allowed, the domestic price will be equal to P_m, domestic consumption and production are, respectively, C_0 and Q_0, and milk imports equal M_0. If an import tariff of t is applied, the domestic price will rise to $P_m(1+t)$, inducing a decline in domestic milk consumption, a rise in production, and a decline in milk imports. It can again be shown that national welfare is reduced by the area of the two shaded triangles. The rise in domestic price induces consumers to shift their expenditure to other goods whose contribution to consumer's utility is somewhat less than that which would have been offered by milk at the initially lower price. Similarly, producers are induced to shift more resources into milk production, which provides less benefit than could be achieved had these resources been used to produce goods in which this country was innately more productive.

Note that the import tariff which increases the domestic price again provides government revenue. It also increases producer incomes by an amount approximately equal to the area abdc, but reduces consumer welfare by an amount approximately equal to area aefc, which is larger than the gains achieved by producers and the government. Thus, milk producers are likely to favour an import tariff even if it decreases national welfare.

Although the conventional economic analysis laid out above indicates that the distortions introduced by a tariff on milk imports is likely to reduce domestic welfare, this analysis may be qualified in important ways. For example, if the international import price is temporarily low due to "dumping," but is expected to return quickly to a higher, normal level, a developing country may find it attractive to impose a countervailing import tariff to ensure that the domestic milk industry does not suffer irreversible harm during the period of low prices. An argument can be made that the gains from importing cheap milk during the interim period will be smaller than the long term losses inflicted on the dairy sector. Whether this is true depends on the relative costs and benefits of the alternative actions, which is an empirical issue likely to vary significantly from one country to another.

Similarly, if the country has potential to significantly increase its milk production efficiency in the long run, a transitory import tariff may permit, if combined with other actions like a strong research program, the realization of such potential. This "infant industry" argument for a tariff is often advanced for the manufacturing sector. A problem, however, is that the alleged increased efficiencies sought are often not achieved and the short run tariff becomes a long run tariff, protected by vested interests. If so, although producers gain from a higher domestic price, the loss by domestic consumers will be larger.

In a number of countries, including India and Indonesia, domestic dairy development has been encouraged by significant import tariffs and/or the receipt of concessionary milk imports which were sold by the government at market prices. The revenues thus obtained being used to subsidize the development of the domestic dairy sector (Alderman). Although such efforts have achieved significant improvements in the quantity and quality of milk produced, in each case, consumers have had to pay higher prices of milk than they would have had to pay had milk been imported at cheaper prices. Thus, if cheap milk was desired for consumers, the policy of applying tariffs has achieved the reverse, at least in the short run. However, since higher income consumers account for the bulk of milk consumption, such policies might have distributional justification in some cases.

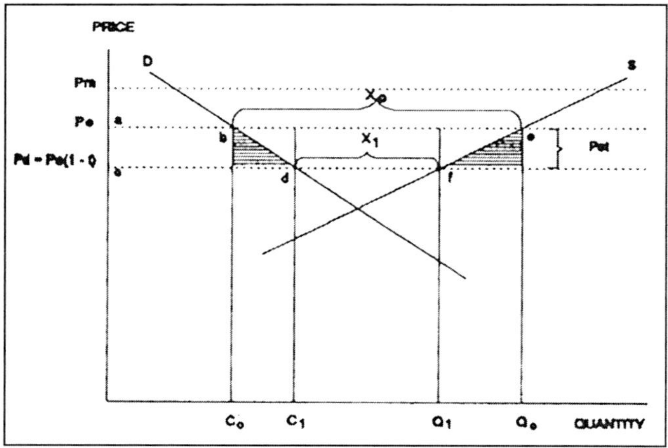

Fig. The Welfare Effects of a Beef Export Tax

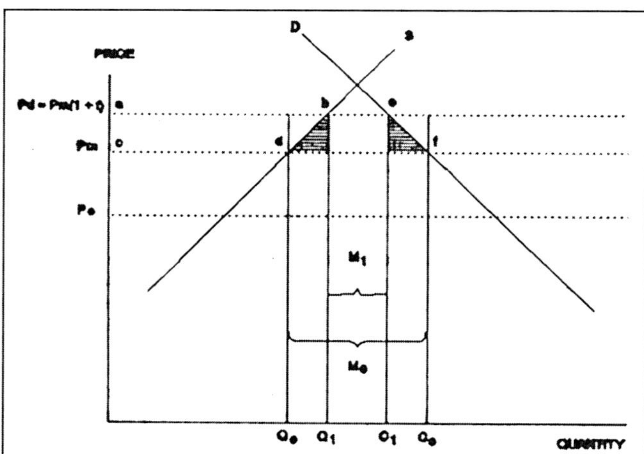

Fig. The Welfare Effects of a Milk Import Tariff

Price intervention via export taxes, import tariffs, and fixed prices are common in developing countries. This chapter has attempted to show that they can cause significant economic damage, even where they increase livestock output. The appropriate policy goal ought to be to optimize total economic output, not the output of any specific product - even livestock!

Several other types of economic interventions warrant mention. One is exchange rate overvaluation, which occurs when the domestic currency purchases more foreign currency than it would at an "equilibrium" in the balance of payments. Exchange rate overvaluation generally occurs as the result of economic development policies which restrict imports to encourage industrial development. The reduction of imports reduces the demand for foreign currency, which in turn leads to a stronger domestic currency. The stronger domestic currency directly reduces the domestic value of all traded commodities. For exported goods, for example, it has an effect similar to that of an export tax. It will do the same for imported products if these are not protected by tariffs.

Empirical studies of policies implemented in many countries suggest that the indirect effects of currency overvaluation have probably been more damaging to the livestock sector than have the direct interventions like price fixing and export taxes, which are more visible. Rarely have agricultural sector spokespersons fully appreciated the damaging effects of currency overvaluation. Even when they have, however, they have usually had little influence on such policies. Although agricultural interests usually influence the policies which directly affect the agricultural sector, the influence of such interests on macroeconomic policies, i.e., the exchange rate, is usually limited.

Government have occasionally sought to prohibit certain types of beef slaughter. For example, in Colombia in the mid-1960s and in Chile in the early 1970s, slaughter of heifers and cows that were deemed still "productive" was prohibited in an attempt to force expansion of the breeding herd and thereby output. This measure led to increased clandestine slaughter of females. Similarly, in Indonesia, the slaughter of younger bulls was prohibited with the intent of building up the livestock herd. It can be shown, however, that the constraints imposed on producers by these prohibitions reduced overall livestock profitability by impeding the producer from taking what appeared the most profitable action. Farmers generally perceive better the appropriate use of livestock capital than do governments. In each of these cases, when farmers slaughter animals they do not generally cease to own livestock assets, but simply shift from ownership of an older less efficient animal to a younger more efficient animal. The government's effort to dictate the appropriate use of animals thus doubtlessly led to a long-run decline in the livestock herd and in output rather than the increases sought.

In most developing countries, industrial protection policies (import tariffs and quotas) has increased the cost of inputs important to livestock production, such as machinery and equipment, veterinary supplies, minerals supplements, fertilizer, and fencing materials are usually high relative to costs in developed countries. The lack of competition among importers of many brand name products is also a factor in high input costs. Such costs inhibit the development and adoption of improved livestock technology. Given the tendency of government market intervention to cause harm, most economists argue that governments should minimize such interventions. However, there are other non-market interventions, like those in public health, veterinary services, research and extension and the provision of infrastructure, which governments must undertake.

LIVESTOCK RESEARCH, INFRASTRUCTURE, AND OTHER GOVERNMENT INTERVENTIONS.

It has become increasingly clear that achieving higher rates of livestock output associated with higher incomes for producers and lower prices and higher quality products for consumers, depends on obtaining improved technology. Probably no aspect of government policy is as important as research for the future of livestock production in developing countries. In several countries where efforts have been made to develop a domestic dairy industry, via import protection, the increase in output has probably come more from technological diffusion than from the stimulus of higher prices, *per se*. For example, in India, the increasing use of crop by-products as feed concentrates, along with the introduction of crossbred dairy cattle having higher potential yields, has resulted in a rapid increase in milk output per animal. Similarly in Colombia and Brazil, the spread of pasture raised crossbred dual purpose cattle has led to rising output through a combination of increased herd size and higher output per animal. In each case, the output expansion has occurred while prices have been roughly constant or declining. Further, as domestic output has risen, reaching self-sufficiency levels, policy makers have been confronted with the possible need to export milk surpluses at world prices which are substantially lower than either the protected domestic price or, even, the milk import price.

Even when basic technology is available in developed countries, a strong domestic research capability is essential to identify and adapt promising technologies to local conditions. Unfortunately, most livestock research and development institutions are weak and a major effort is needed to strengthen them, especially, the number of qualified staff.

An institutional environment which allows farmers to a) market their produce for higher prices, b) obtain inputs at lower cost, and c) obtain information necessary for decision marking more rapidly and effectively is critical to livestock development. Infrastructure is especially important where timely access to inputs is essential and where livestock product must be marketed quickly. Milk production is such an example and depends heavily on an efficient the establishment of a marketing and distribution system, including milk collection, transport, and processing. The diffusion of new technologies to farmers, e.g. artificial insemination and forage technologies, has often been achieved through the same systems. Such efforts have been carried out efficiently by private firms and cooperatives, though rarely by public firms. Livestock diseases impose significant economic losses and can present a threat to human health. Such losses and threats can be economically reduced by improved animal health programmes. However, programmes such as vaccination, sanitation and inspection programs require a deep commitment to implementation from producers and governments and often require regional cooperation.

THE EFFECT OF SOCIAL AND POLITICAL FACTORS.

Social factors can influence the use of livestock such influences are often obvious, for example, religious objections to the consumption of pork in Muslim countries and beef in Hindu countries. In other cases, however, the effects of social influences are less pronounced though still important. For example, the attribution of prestige to ownership of cattle may increase the cattle's total value but lead to other tangible products like meat and milk. In general, such effects seem relatively small in magnitude.

In much of Asia cattle, utilizing locally available residues, are used mainly for animal traction on small farms. In India, for example, cattle are fed principally on wheat and rice straws which account for approximately half the plant's total gross energy. This energy would be largely wasted if it were not consumed by ruminants and converted into traction for ploughing and manure.

Due to religious beliefs beef in India is of little value (beef is consumed by the Moslem minority in some areas). Nonetheless, farmers require some means to cultivate and most farms are too small to justify mechanization and produce forage. The value from draught and manure must therefore be sufficient to justify maintaining cattle and this in fact the case where livestock are kept. Incorporation of manure into fields and garden areas, to improve soil fertility and structure, is often essential to maintaining agricultural production. It can be shown that the social constraint on the eating of beef, by diminishing the value of beef which is a necessary by-product of milk, traction, and manure production, makes cattle a less profitable investment. Further, with no incentive to slaughter the animal, the tendency is for producers to simply abandon their animals when their economic life animals has ended. Such animals are a general nuisance and harm other farmers who need to protect their crops and from consuming valuable common grazing.

THE SUSTAINABILITY OF LIVESTOCK PRODUCTION SYSTEMS.

Livestock are usually helpful in sustaining agricultural production. However, there are cases where livestock development has had disastrous environmental consequences. For example, clearing of the tropical forests in Central America and the Amazon during the last two decades, these developments has been sharply criticized for their ecological and sociological damage. Most criticism has focused on a) the destruction of irreplaceable genetic materials, b) tendency for pasture to rapidly diminish in productivity because of loss of soil fertility, leaving the fragile soils vulnerable to compaction and erosion, c) the displacement of indigenous peoples and small farmers by land speculators who have used cattle ranching as a mechanism for obtaining and controlling large tracts of land, and d) the threat to the environment from destruction of oxygen producing trees.

Livestock development, *per se,* in most of the Amazon basin is not very profitable at current prices. Nonetheless, government incentives in Brazil have affected livestock development and, more dramatically, Amazon settlement and deforestation (Binswanger). Income tax credits and subsidized interest rates on loans for livestock development, along with grants of land on favourable terms to individuals engaged in livestock development, have given substantial private incentives for livestock development in rainforest areas. This is one of the most dramatic examples of a case where government policy is the primary cause of an unsustainable agricultural system. Although some rainforest destruction would remain even if government policies were fully neutral, due to the pressure of spontaneous colonization by poor farmers but the areas affected would be much smaller. The damage caused by such settlements is a more difficult problem. Achieving a sustainable system in such situations will require development of either improved technologies or, more likely, the exclusion of settlers.

Developed countries contribute to environmental degradation to a far greater extent than does either Brazil or other developing countries. However, that others have and are destroying irreplaceable assets is poor justification for continuing with equally bad policies in the developing countries. The intent should be to make the best use possible of the available resources, in all regions.

Overgrazing on semi-arid and arid lands, leading to range degradation, is another case of an unsustainable livestock system. Pastoralism is a practice that utilizes extensive rangelands where rainfall is low and highly variable, making settled agriculture and/or livestock production extremely risky. The principal production risk for a specific area is that no rain will fall. However, cattle can be herded to areas in which rain has fallen and pastures are available.

The need to access a large area of land in order to ensure sufficient pasture is an important reason for the evolution of "common" range systems. In such systems, a group of pastoralists share land, with all being able to move about with their herds in search of the best forage. If the system is to work well, pastoralists must have a well defined membership group with clear (albeit sometimes complex) rules of access to pasture and water. If group membership and/or the rules of access become unclear or ineffective, particularly as when population or economic pressures encourage greater use of the range, the system may tend toward an "open" access system in which no limits are placed on the number of herders (and animals) using the land. In this case, the economic value of pasture is likely to be severely diminished or lost altogether.

In most areas, pastoralism is probably more productive in terms of the value of total output of beef and milk per hectare of land than is cattle ranching. The pastoralists utilize much more labour and extract a larger number of joint products for direct use, especially milk. Beef is produced from cull animals, both steers and cows, but accounts for a relatively small proportion of output. Pastoralists often barter milk and beef with agriculturalists for grain, which is a cheaper source of energy. Manure is used for fuel, and is also left on the fields of agriculturalists during seasonal migrations into settled areas. Agriculturalists sometimes pay herders to graze their animals overnight on their fields.

In systems in which land is communally owned, livestock ownership provides usufruct rights to land which are otherwise lost. Mechanisms are needed in such systems to ensure that all individuals having grazing access also have livestock. In pastoralists systems, livestock ownership traditionally belonged to kinship groups which used force to maintain their hegemony over a particular region. Complex societal rules and livestock exchanges existed within such groups to ensure that individuals who lost their animals to disaster, such as drought or disease, could reconstitute their herd. Such mechanisms have been breaking down in recent years, largely because pastoralist populations have gradually expanded while rangelands have been lost to the spread of sedentary agriculture. Under these conditions, the average herd has been shrinking whilst the aggregate number of animals grazed has been growing.

Traditional mechanisms have proved insufficient to reconstitute the herds of many individuals following disaster. Wealthy individuals, frequently located in urban areas and able to better diversify risks through other economic activities, are accumulating animals and hiring others to herd them. Gradually, as a higher proportion of total herds are owned by such individuals who seek a more marketable output, greater emphasis is being placed on beef production.

Under such pressure, there is fear that overgrazing is increasing. It appears, nonetheless, that the main effect of overgrazing has been an increase in the periodic herd losses suffered from drought, rather than a decrease in range quality. The range generally seems to have substantial resiliency, recovering more rapidly than the herd. However, such a system is truly sustainable, except perhaps at a low average and highly variable yield. A greater problem is the lack of incentives created by the common range system for the development of any productivity increasing technologies. Such technologies would be privately unprofitable within a situation where animal nutrition is not under the herder's control. Without a shift toward greater control of land, there is little possibility of increasing ruminant livestock output in this area.

The primary problem faced by common ranges is the inequities which are likely to be created as common lands are converted into lands with an increasing degree of private control. However real, important and difficult these issues are, it is not an adequate justification for retaining the current system. Historically, pastoralist groups have fought for specific areas, thus deciding "property rights." In recent years, the increased value of livestock output has led to the gradual emergence of more private land rights in countries as diverse as Somalia and Botswana. Kenya has successfully privatized its land with beneficial results. Clarification of such land rights is the primary factor needed to reduce overgrazing. However, that will not in itself reduce the increasing pressures on land which are stemming from higher population and which lead to deforestation and cultivation of marginal land. Again, the best solution seems to require the development of improved technologies which are more productive and sustainable combined with the education of producers to use them.

Feed availability, disease, climate, social and political forces, as well as economic incentives all influence the pattern of livestock use which emerge in developing countries. Thus, one of the main responsibilities of policy makers is to ensure that economic markets work well so that livestock producers receive appropriate signals regarding resource allocation. Similarly, the sensitivity of the mix of outputs to economic and technical factors indicates that, when formulating livestock development strategies, it is important to have a clear understanding both of the various production constraints and also of the demand the for different products.

There are cases where government intervention in markets is justified i.e. the protection of domestic producers from international dumping or to initiate long term industry development. Unfortunately, development often does not always occur even when the government has intervened to obtain higher livestock prices. Sometimes this is due to the intervention appears short term and the resulting uncertainty mitigates against development which were in fact sought. More generally, the government does not make the concerted effort through related research to develop the technologies required to increase productivity, or the requisite political stability or land rights are not present. As a general, though not precise, rule of thumb, livestock development in most countries will make the greatest contribution to domestic welfare if the government establishes free markets for livestock products and inputs and strives to develop complementary research, infrastructure and animal health programs. Research, education, and non-distorted market signals seem generally to be the best guarantee of achieving sustainable livestock production.

PRACTICAL TECHNOLOGIES TO OPTIMISE FEED RESOURCE UTILISATION

Production from a herd or flock of ruminants is a result of the interactions of environment, the animals nutrition, and its genotype.

Individual productivity from most ruminants in all developing countries is low. The reasons for this are complex but in order of priority appear to be:

- The imbalance of nutrients that arise from digestion of the available forage resources when these are fed without supplements,
- The incidence of disease/parasitism, and
- The often harsh climatic conditions.

Genotype is over-emphasised as a primary constraint as poor nutrition has an overwhelming effect. Resistance to disease and high temperatures of particular breeds is however an important overall consideration.

Recent nutritional research has demonstrated the possibility of very large increases in animal production that can be achieved by small alterations to the feed base. As these increases have also been achieved at the farm level without alteration to the other management practices, it demonstrates the large impact potential of the feeding strategies in the present environment. Increased production of meat/milk with better body condition of animals also lifts lifetime reproduction rates. The lowering of the age at puberty of heifers and a decrease in the intercalving interval in cows have probably the greatest effect on the overall level of production.

Increased efficiency of utilisation of forage by the animals together with improved reproduction rates have demonstrated that production can be increased by up to five fold without changing the basal feed resources. This has been achieved, by providing the critical catalytic nutrients that are deficient in the diets and by balancing availability of nutrients closer to requirements.

In general, the supplements required are urea/minerals and a source of bypass protein. In many countries these supplements are already available, in others, there is a need to manipulate these resources, to provide them in the correct amounts and in the appropriate form.

In many grazing areas the basic resources are not available locally. Research and development is needed to produce them economically at the centres of ruminant population densities. In the rangelands, particularly in the semi arid areas, tree forages, seeds and pods represent by far the greatest potential source of protein meals.

AVAILABLE FEED RESOURCES FOR RUMINANTS IN DEVELOPING COUNTRIES

Throughout the last thirty years the expansion of crop and livestock production in developing countries have more than doubled but the increase in demand for food has been even greater, leading to an increase of food imports by approximately 10% per year. This situation is expected to remain at the same rate for the foreseeable future.

The increased production of animal products in developing countries has been largely through increased animal numbers, while production per animal has been static or increased to a minor extent over a long period of time. The demand for food for humans in these countries is likely to increase and, since cropping land is almost fully utilised at the present time, it appears that cultivated pasture is likely to become scarcer in most parts of the developing world.

The ruminants niche is likely to remain as a utilizer of carbohydrate biomass which is not digested by intestinal enzymes and, therefore requires fermentative digestion, and which cannot be used extensively by monogastric animals (i.e. forage, crop residues etc.). The ability of ruminants to convert otherwise waste biomass into meat, milk and other products and to accomplish work suggests that they will endure into the foreseeable future. The ever increasing pressure on land for crop production suggests, however, that they will have to continue to do this utilizing crop residues, industrial by-products and pastures from relatively infertile rangelands. The common characteristics of such feeds are low digestibility, low protein content and a low mineral component.

ANIMAL PRODUCTIVITY

Animal Productivity from Available Feed Resources

Research over the last twenty years clearly indicates that it has been a popular misconception that low productivity of ruminants in developing countries is a result of low energy density of the available forages (i.e. low digestibility). This concept, often repeated in reviews, even up to the present time, is misleading. There is now abundant evidence that low productivity stems from an inefficient utilisation of the feed because of deficiencies in the diet. These deficiencies are of nutrients critical to the well being of the microbes which ferment or digest the feed, and nutrients required to balance the products of digestion to requirements. This has been considered in detail recently by Leng (1991), who suggested that because an inefficient utilisation of nutrients increases metabolic heat, that the often low intakes of poor quality forage of ruminants in the tropics (where most developing countries exist) is imposed by a combination of climatic and metabolic heat stress. Correction of a nutrient imbalance by feeding a bypass protein often (but not always) increases the intake of poor quality forages to a constant intake of around 80-100 $g/kg^{0.75}/d$. Tropical conditions impose this particular feed intake constraint on ruminants for a considerable period of a feed year and it is rarely seen in temperate areas. On the other hand cattle in the tropics require less feed for maintenance, if they do not have to combat cold stress, and if they can process these extra nutrients they can be more efficient than animals on the same feed in a cold climate. To process the extra nutrients however, they need extra protein and thus the requirements for amino acids are higher in cattle in the tropics then in animals on the same feed in cold environments.

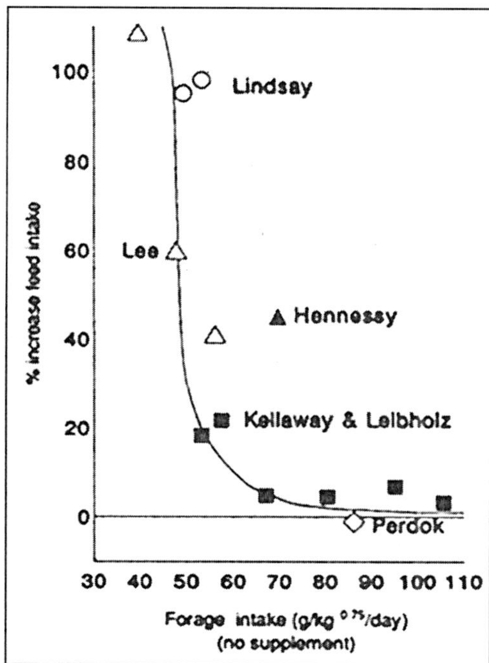

Fig. Intake of low digestibility forages by cattle either unsupplemented or supplemented with bypass protein or bypass protein and urea.

Productivity Levels Achievable by Ruminants on Low Quality Forage Based Diets

The rationale and concepts on which the following discussions will be centred have been reviewed by Preston and Leng (1987) and Leng (1989,1991). The basic concepts are as follows. Ruminants fed low quality forages require supplementation with critically deficient nutrients to optimise productivity. The supplements that are required:

- Correct any nutrient deficiencies for the rumen microbes, and
- They balance the ratios of protein (absorbed amino acid) to the energy (VFA) that arises from the fermentative digestion in the rumen so that it corresponds to the animals requirement.

The nature of anaerobic microbial fermentation in the rumen indicates that microbial cell growth in the rumen (which supplies the protein {P} to the animal) relative to the production of VFAs (E) (the major source of oxidisable substrate for ATP generation), will be extremely low in nutrient deficient rumen medium or digesta. This means that a sub-optimal level of any nutrient for microbial growth in the rumen will result in low protein to energy (P/E) ratio in the nutrients absorbed. Ensuring a nutrient non-limited microbial digestion in the rumen by supplementation automatically improves the P/E ratio in the nutrients available to the animal. Feeding a protein meal in which the protein has been made insoluble or otherwise not attacked by rumen microbes, is a further major method for adjusting the P/E ratio upwards.

The ratio of microbial proteins to VFA produced and the effects of supplementation in a steer given 4kg of organic matter which is completely digested in the rumen.

The reason for discussing these theoretical calculations at this point is to emphasise the large differences in protein to energy (from 12 to 50) in the nutrients absorbed by ruminants, fed unbalanced diets and diets balanced with supplements. It also emphasises that on a diet high in bypass protein the rumen microbes need not be highly efficient. To obtain the higher P/E ratio without supplementation, however, to directly improve rumen condition, more bypass protein is needed.

Table. The effects on P/E ratio in the nutrients absorbed of supplementation with a bypass protein to cattle with a poor or optimised (i.e. supplemented) microbial milieu in the rumen. The values are calculated for a steer digesting 4 kg DM in the rumen

Rumen Environment Protein p(g/d)	Protein by Pass p(g prot/d)	Microbial Cells Produced	Protein Microbial p(g/d)	VFA Produced p(MJ)	P/E* p (g.MJ)
Poor	0	830	500	41	12
Optimised	0	1680	1010	30	33
Poor**	400	830	500	41	22
Optimised	400	1680	1010	30	47

Note:

* Microbial protein plus dietary protein to VFA energy.

** Although the rumen environment is deemed not to change through the addition of protein meal, in fact it will have been improved but may not be optimised to the extent it would by feeding a molasses/urea block. P/E ratio here is underestimated.

It is the relationship of P/E with the efficiency of feed utilisation that has a very large effect on growth, milk yield and reproductive performance. The levels of production achieved when P/E is increased have been greatly superior to that predicted from present day feeding standards based on the metabolisable energy of a feed.

FEEDING STANDARDS AND FEED EVALUATION

Most forages consumed by livestock in developing countries have a low digestibility which rarely exceeds 55% and is mostly in the range of 40-45%. The calculated metabolisable energy in the dry matter (M/D) thus ranges from 7.5 down to 4.8. Feeding standards indicate that feeds with a metabolisable energy content of 7.5 will support growth rates of cattle of approximately 2 g/ MJ of M/E intake. On a forage at the lowest level of ME, cattle would be in negative energy balance.

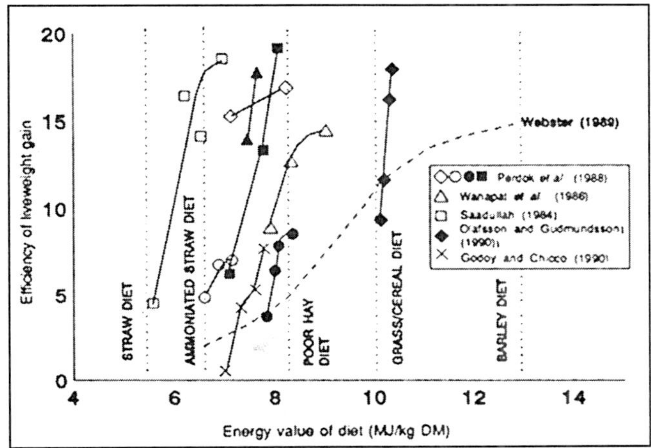

Fig. Schematic relationship between diet quality (metabolisable energy/kg dry matter) and food conversion efficiency (g liveweight gain/MJ ME) from Webster, 1989.

Contrast this with results of supplementary feeding trials based on balancing the nutrition of animals with urea/minerals and bypass protein, where cattle growth rates equivalent to 18 g/MJ of M/E intake have been achieved in cattle fed straw. Obviously the presently accepted feeding standards have been very misleading and can not be used as a means of predicting animal performance. Of vital importance however, is that the application of the concept of balanced nutrition can improve animal growth by 2–3 fold and the efficiency of animal growth by as much as six fold over previous estimates (a range of 2–10 fold). In addition it also shows that although growth rates of cattle are below those on grain based diets cattle on forage based diets can be as efficient in converting feed to liveweight gain.

Low productivity of ruminant livestock has been accepted in developing countries as an inevitable result of the poor feed base and a low feed conversion efficiency. The concept being that there is a large heat production (energy requirement) associated with the ingestion and movement of digesta along the tract in animals fed on forages as compared to concentrates. This conclusion is clearly contrary to the conclusions of Leng (1991) and the concept of balanced nutrition presented here.

However, poor growth rates are associated with a slow maturity and, on poor quality roughage diets, cattle reach puberty at four to five years of age and often have a calving interval of 2 years. Recent observations on balancing the nutrition of cattle on such forages indicate that it is possible to reduce age at puberty by one to two years and potentially possible to support a calf every year on these same feed resources. Management considerations, however, suggested that 15 months calving interval is more likely. The flow-on effects of improved reproduction are: increased percentage of a herd in production and an increased offtake of animals. The flow-on from improved reproduction outstrips the direct effects on immediate liveweight gain or milk yield.

APPLICATION OF SUPPLEMENTATION TO BALANCE P/E RATIOS

Even though the principles of feeding bypass protein to improve productivity have been known for many years, application has been slow and unspectacular, particularly in the developing countries. The application has been slowed by:

- The inability of research scientists to communicate and be believed by applied technologists;
- The desire by many scientists to stand by the feeding standards that have been promulgated for twenty years and which appear to be totally inappropriate for most feeding systems;
- The controversies surrounding the principle mechanisms ofaction of protein supplementation which has clouded the major issues.

However, wide scale application of the use of supplements high in bypass protein have occurred in India through the initiatives of The National Dairy Board of India (NDDB). For the same reasons as given above, progress was initially slow (development started in 1980) but it is now accelerating at a pace which should see most feed mills in India dedicated to the production of bypass protein supplements in the next five years. At the present time approximately 100 MT of bypass protein feed is being fed daily to their dairy animals by small village farmers. In many situations, this is coupled with the use of a molasses urea block, particularly by the more advanced farmers.

After considerable experimentation and village testing of a bypass protein supplements, the management of the cooperative feed mill in Kedah district in India decided to convert from the production of concentrates based on traditional concepts to a bypass protein concentrate (30% CP) composed of locally available protein meals plus 10% grain and approximately 10% molasses. After pelleting, the protein in the concentrate was found to be 75% insoluble in buffer solution.

The marketing strategy was to widely advertise the new feed concentrate with a simple statement that village farmers should —— "feed to their dairy animals half the weight of the usual concentrate and this would double milk production". As the feed was about a third more expensive, there was some considerable opposition to its introduction. Nevertheless, the feed mill (capacity 100 MT/day) was converted to the production of new feeds on December 1st 1988. The feed was purchased, somewhat reluctantly by some of the farmers, but all opposition to its supply appeared to have been overcome by the middle of the next hot dry period (April-June) when milk yields were considerably above previous years with only half the weight of supplement. The collection of milk within the dairy co-operative, relative to the feeding of the traditional and new feed supplement for the previous five years and for the twelve months since conversion to the new feeding systems. Whilst a number of changes have occurred in the area which could contribute to the increased milk collected, from the research carried out by NDDB prior to the change over the responses in milk yield are in line with observed increases in milk collected.

The effects of the changed feeding appear to be a 30–50% increase in milk production, from Dec. 1st. 1988 to Dec. 1st. 1989 - a further similar increase in production is apparent in 1989-1990 (unpublished observation). The further increase probably represents the flow-on effect that would come from increased reproduction rate that should have resulted from the new feeding systems.

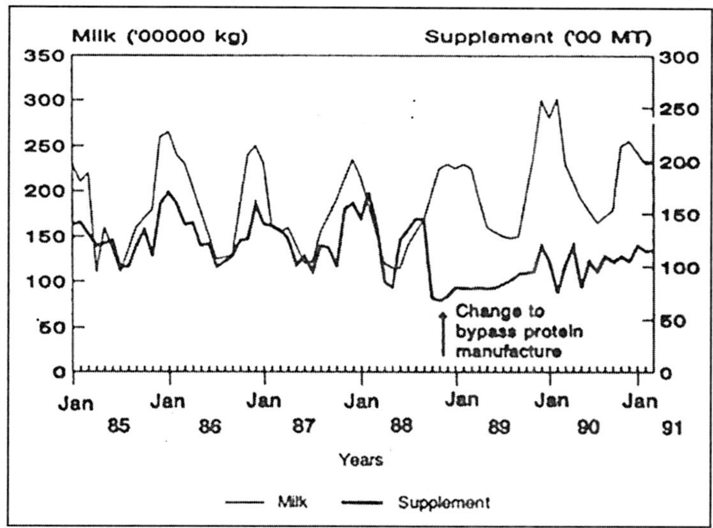

Fig. Milk collection records and the sale of supplements in for the Co-operative in the Kedah district of India when supplements were compounded on traditional concepts (1985–1987) and following (1st December, 1988) their replacement with a 30% C.P. bypass protein pellet (records provided by the NDDB of India).

Impact of New Feeding Systems on Milch Herd Composition

Undoubtedly the new feeding systems, combined with disease control and better management, is facilitating the introduction of cows with a higher genetic potential for milk production. The apparent lowering of heat stress through the balanced approach to nutrition has also allowed Friesians (the mothers of bulls that will be used for cross breeding purposes) and selected buffaloes, to yield milk well above the average reported for the developed countries. The yields of imported Friesian cows from Germany fed on all roughage diets supplemented with a molasses/urea block and a bypass protein (recommended at 350 g/litre of milk) at Anand averaged over 5,000 litres/305 day lactation in 1988–89 with individual milk yields of over 6,000 litres/305 day. These milk yields are achieved in an area of India considered to have one of it's hottest climates. Higher yields were observed with Friesians managed by village farmers in a more temperate area of India (Bangalore).

PRACTICAL METHODS FOR BALANCING NUTRITION

Provision of Minerals

It is not practical to identify the major critical micro and macro minerals in a basal roughage diet, as these will vary from site to site and year to year. The practical approach is one of 'rules of thumb' that provide a best bet or 'shot-gun mixture' of minerals as economically as possible. A concentrated plant extract, such as, molasses provides such mixtures and can be fortified for specific areas where local knowledge points to specific deficiencies. In this respect, molasses (both sugar cane and beet) and concentrated palm oil sludge offer useful sources of these minerals. They are also quite palatable to livestock and are useful in hiding less palatable nutrient sources in supplements. Mineral salt mixtures are commercially available in most countries. They usually have a high content of salt and only minor quantities of the trace elements. In practice, fortified materials, such as molasses, will be superior to commercial mixtures as they present a greater coverage of all the minerals required, also thy are a valuable source of other nutrients (e.g. B vitamins) and a small amount fermentable energy.

Supplying the Rumen Microbes with Ammonia/Urea

The other requirement is for a non-protein nitrogen source for the rumen microbes, usually urea. Urea is usually administered together with the minerals and its concentration in such mixtures is controlled by safety and ease of incorporation and, therefore, rarely exceeds 10–15% of such mixtures. However, this is usually sufficient to allow an intake of between 50 and 100 g of urea by cattle from a molasses/urea block which is sufficient to balance ammonia in the rumen of cattle on a low N roughage diet. Results from India suggest that mineral/urea mixtures in, for example, molasses multi-nutrient blocks, are best given *ad libitum* allowing the animal some degree of selection and there are indications that the animal will learn to control the intake of urea to an optimal level.

Bypass Protein Supplements

Providing bypass protein to cattle under small farmer management is often difficult and, at times, is too expensive. There is often little information on the locally available protein sources, particularly the level of protection of the protein in the rumen. As a rule-of-thumb, solvent extracted oilseed cakes, fish meal that has been flame dried (but not sundried or fish silage) and protein sources that have been heat treated, have considerable protection from rumen degradation. The degree of protection is enhanced by pelleting the protein meal in the presence of free glucose or fructose, (as occurs in molasses) when a mild browning reaction occurs (unpublished observations).

The turning over to supply bypass protein rather than concentrate for supplementing the cattle of small farmers, the National Dairy Development Board of India converted the existing feed mills to the production of high protein supplements. Wherever possible, solvent extracted oilseed meals form the basis of the supplement. However, the protein ingredients have to be purchased on the open market and leastcost formulation is desirable because of the size of the production unit (50–100 MT/d.). The high degree of protection of these protein feeds is achieved by ensuring a high proportion of solvent extracted meal and also by heat pelleting with 8% molasses in the mixture.

IDENTIFICATION OF PROTEIN SOURCES

India is fortunate in having large amounts of crop residues high in protein, most of which have a fair degree of protection brought about by the processing methods. These materials are convenient to use in existing feed mills with the available equipment (i.e. grinders, mixers and pelleters) and the pelleted supplement is readily accepted because of a well developed marketing strategy. In many other countries, particularly in extensive grasslands or savannahs, where major constraints to production of cattle are essentially the same as those for cattle fed crop residues, protein sources may not be readily available, or the sources not so obvious or easily obtainable. Most legume forages, legume seeds, edible tree leaves, seed pods and seeds that are available in these areas, contain highly soluble proteins which is easily fermented in the rumen. These, when used as supplements, only provide ammonia and minerals (they have for example 0.5-1% phosphorus). Fed without processing, because of their influence in the rumen they generally increase production of cattle on a basal diet of low protein roughage, but they provide little bypass protein unless they are fed as a large proportion of the total diet. Research results in Colombia fit into the general concept where supplementation of cattle on green *Brachiaria decumbens* pasture with either urea/molasses or the foliage of the fodder tree *Glyricidia*, increased production to the same extent.

Table. The effects of feed supplementation on livestock gain of cattle (6 per group) grazing on green *Brachiaria decumbens* pastures in the wet season (with mineral supplements) with liquid molasses/urea 10% or *Glyricidia* foliage.

Treatment	Rumen Ammonia p(gN/1)	Initial p Wt (kg)	Final p WT (kg)	LW t Gain p(g/d)
No supplement	50	194	244	580
+ *Glyricidia*	170	204	266	717
+ Molasses/Urea	250	203	269	751

Protein that is fermented in the rumen yields 80–100 g of microbial protein/ kg of protein consumed. Whereas carbohydrate yields 180–200 g of microbial protein/kg of carbohydrate. On a high nitrogen diet, a significant supplement of soluble protein could in fact imbalance the P/E ratio because of the low microbial growth efficiency. This is a possible explanation for recent results reported from Iceland where fish silage and fish meal were compared as protein supplements to 'high quality' grass silage fed to young cattle. Fish silage, with the same N content as a fish meal supplement, depressed liveweight gain by 90g/day whereas a fish meal supplement improved liveweight gain by 360 g/ day at the same N intake.

Table. The effect of supplementation of a basal grass silage * diet with fish meal or fish silage on liveweight gain and efficiency of feed conversion.

Supplement	Intake (kg/d) Supplement		Silage DM	LWt Gain p(kg/d)	Calculate d ME Intake	Efficiency (g LWt gain/MJ ME)
	DM	CP				
	.200	.123	4.6	.838	50	17
Fish	.400	.246	4.5	.914	51	18
silage	.400	.138	3.4	.358	40	9

Note:

* Early cut, precision chopped and preserved with formic acid, DM In vitro DMD. = 73% Calculated M/D = 10.4. CP = 16.8%. Cattle were Galloway/Icelandic cross - 6 months (8/group treatment), 160–205 kg LWt at beginning of experiment.

The above discussion defines and highlights two potential strategies to provide the two types of supplement required to optimise the efficiency of feed utilisation of cattle in areas with scarce resources of nitrogen or protein. The strategies must be to find forages/trees/tree seeds or pods that are high in protein and minerals. To then use this material in small supplements to provide for either the rumen with soluble protein and minerals or bypass protein after treatment to protect the protein. These uses can be combined with either a molasses/urea block and/or a source of locally available bypass protein.

Processing of Local Protein Resources to Provide Bypass Protein

The protein and render them non-fermentable in the rumen but allow them to retain digestibility in the intestines. In general, these fall into chemical treatment with agents that cross link with amino acids on the protein chain and include formaldehydes and aldehydes, tannins and simple sugars such as glucose and xylose. These reactions often require the protein source to be heated during processing. Heat alone will often denature the protein to effect protection. For instance, Goering and Waldo (1974) found significant effects of the temperature used to dry lucerne on the subsequent animal production from that lucerne. In general, the higher the temperature of drying the greater the retention of protein by the animal. More recently, Lewis *et al.* (1988) have demonstrated that mild heat with a small amount of xylose is very effective in protecting soya bean meal protein. Xylose can be readily produced by acid hydrolysis of many fibrous materials including bagasse and cottonseed hulls.

Table. Effects on liveweight gain of cattle supplementing a basal forage/concentrate based diet with soyabean meal or soyabean meal treated with sulfite liquor (SL) at 200°F for 2 hours.

	LWt gain p(g/d)
No supplement	591
+7% soyabean	673
+9% soyabean + 10% SL	823
+8% soyabean + 5% SL	841

Little work has been done with forage proteins, although there are indications that the presence of low levels of tannins in the forage (e.g. leucaena) does afford some protection of its protein.

There is a great need to research methods for protecting the protein of forages, tree leaves, seeds and pods from local resources, as these are the proteins that are potentially available in the pasture areas of the world.

Trees as a Source of Protein and Soluble Nitrogen in the Rangelands

A number of authors have pointed to the small percentage of the total pasture biomass that is actually consumed by grazing animals in the extensive arid and semi-arid rangelands. Most tropical grasslands are highly leached and their pastures apparently have low potential to provide ruminants with their required nutrients. The point to be stressed is that 10–30% of pasture biomass is often all that is used by grazing animals. Trees, however, can produce considerable amounts of edible biomass. For example, the tree *Prosopis juliaflora* produces up to 440 kg of edible pods per annum with an average 16% crude protein, (Riveros pers. comm) but the protein is, in all probability, soluble. Compare this with the usable biomass of 250 kg DM from the poor native pastures of South America and the combination of trees and grassland would obviously be a desirable development and synergistic for cattle production.

3

Intensive Animal Farming

Intensive animal farming or industrial livestock production, also called factory farming, is a modern form of intensive farmingthat refers to the keeping of livestock, such as cattle, poultry (including in "battery cages") and fish at higher stocking densities than is usually the case with other forms of animal agriculture—a practice typical in industrial farming by agribusinesses. The main products of this industry are meat, milk and eggs for human consumption. There are issues regarding whether factory farming is sustainable and ethical.

Confinement at high stocking density is one part of a systematic effort to produce the highest output at the lowest cost by relying oneconomies of scale, modern machinery, biotechnology, and global trade. There are differences in the way factory farming techniques are practiced around the world. There is a continuing debate over the benefits, risks and ethical questions of factory farming. The issues include the efficiency of food production; animal welfare; whether it is essential for feeding the growing global population; and the environmental impact (e.g. pollution) and health risks.

HISTORY

The practice of industrial animal agriculture is a relatively recent development in the history of agriculture, and the result of scientific discoveries and technological advances. Innovations in agriculture beginning in the late 19th century generally parallel developments in mass production in other industries that characterized the latter part of theIndustrial Revolution. The discovery of vitamins and their role in animal nutrition, in the first two decades of the 20th century, led to vitamin supplements, which allowed chickens to be raised indoors. The discovery of antibiotics and vaccines facilitated raising livestock in larger numbers by reducing disease. Chemicals developed for use in World War IIgave rise to synthetic pesticides. Developments in shipping networks and technology have made long-distance distribution of agricultural produce feasible.

Agricultural production across the world doubled four times between 1820 and 1975 (1820 to 1920; 1920 to 1950; 1950 to 1965; and 1965 to 1975) to feed a global population of one billion human beings in 1800 and 6.5 billion in 2002. During the same period, the number of people involved in farming dropped as the process became more automated. In the 1930s, 24 percent of the American population worked in agriculture compared to 1.5 percent in 2002; in 1940, each farm worker supplied 11 consumers, whereas in 2002, each worker supplied 90 consumers.

According to the BBC, the era factory farming per se in Britain began in 1947 when a new Agriculture Act granted subsidies to farmers to encourage greater output by introducing new technology, in order to reduce Britain's reliance on imported meat. The United Nations writes that "intensification of animal production was seen as a way of providing food security." In 1960s North America, pigs and cows began to be raised on factory farms. From its American and West European heartland factory farming became globalised in the later years of the 20th century and is still expanding and replacing traditional practices of stock rearing in an increasing number of countries. In 1990 factory farming accounted for 30% of world meat production and by 2005 this had risen to 40%.

CONTEMPORARY ANIMAL PRODUCTION

Factory farms hold large numbers of animals, typically cows, pigs, turkeys, or chickens, often indoors, typically at high densities. The aim of the operation is to produce large quantities of meat, eggs, or milk at the lowest possible cost. Food is supplied in place. Methods employed to maintain health and improve production may include some combination of disinfectants, antimicrobial agents, anthelmintics, hormones and vaccines; protein, mineral and vitamin supplements; frequent health inspections; biosecurity; climate-controlled facilities and other measures. Physical restraints,e.g. fences or creeps, are used to control movement or actions regarded as undesirable. Breeding programs are used to produce animals more suited to the confined conditions and able to provide a consistent food product.

Intensive production of livestock and poultry is widespread in developed nations. For 2002-2003, FAO estimates of industrial production as a percentage of global production were 7 percent for beef and veal, 0.8 percent for sheep and goat meat, 42 percent for pork, and 67 percent for poultry meat. Industrial production was estimated to account for 39 percent of the sum of global production of these meats and 50 percent of total egg production. In the U.S., according to its National Pork Producers Council, 80 million of its 95 million pigs slaughtered each year are reared in industrial settings.

Chickens

In the United States, chickens were raised primarily on family farms until roughly 1960. Originally, the primary value in poultry was eggs, and meat was considered a byproduct of egg production. Its supply was less than the demand, and poultry was expensive. Except in hot weather, eggs can be shipped and stored without refrigeration for some time before going bad; this was important in the days before widespread refrigeration.

Farm flocks tended to be small because the hens largely fed themselves through foraging, with some supplementation of grain, scraps, and waste products from other farm ventures. Such feedstuffs were in limited supply, especially in the winter, and this tended to regulate the size of the farm flocks. Soon after poultry keeping gained the attention of agricultural researchers (around 1896), improvements in nutrition and management made poultry keeping more profitable and businesslike.

Prior to about 1910, chicken was served primarily on special occasions or Sunday dinner. Poultry was shipped live or killed, plucked, and packed on ice (but not eviscerated). The "whole, ready-to-cook broiler" was not popular until the 1950s, when end-to-end refrigeration and sanitary practices gave consumers more confidence. Before this, poultry were often cleaned by the neighborhood butcher, though cleaning poultry at home was a commonplace kitchen skill.

Two kinds of poultry were generally used: broilers or "spring chickens"; young male chickens, a byproduct of the egg industry, which were sold when still young and tender (generally under 3 pounds live weight), and "stewing hens", also a byproduct of the egg industry, which were old hens past their prime for laying.

The major milestone in 20th century poultry production was the discovery of vitamin D, which made it possible to keep chickens in confinement year-round. Before this, chickens did not thrive during the winter (due to lack of sunlight), and egg production, incubation, and meat production in the off-season were all very difficult, making poultry a seasonal and expensive proposition. Year-round production lowered costs, especially for broilers.

At the same time, egg production was increased by scientific breeding. After a few false starts, (such as the Maine Experiment Station's failure at improving egg production) success was shown by Professor Dryden at the Oregon Experiment Station.

Improvements in production and quality were accompanied by lower labor requirements. In the 1930s through the early 1950s, 1,500 hens was considered to be a full-time job for a farm family. In the late 1950s, egg prices had fallen so dramatically that farmers typically tripled the number of hens they kept, putting three hens into what had been a single-bird cage or converting their floor-confinement houses from a single deck of roosts to triple-decker roosts. Not long after this, prices fell still further and large numbers of egg farmers left the business.

Robert Plamondon reports that the last family chicken farm in his part of Oregon, Rex Farms, had 30,000 layers and survived into the 1990s. But the standard laying house of the current operators is around 125,000 hens. This fall in profitability was accompanied by a general fall in prices to the consumer, allowing poultry and eggs to lose their status as luxury foods. The vertical integration of the egg and poultry industries was a late development, occurring after all the major technological changes had been in place for years (including the development of modern broiler rearing techniques, the adoption of the Cornish Cross broiler, the use of laying cages, etc.).

By the late 1950s, poultry production had changed dramatically. Large farms and packing plants could grow birds by the tens of thousands. Chickens could be sent toslaughterhouses for butchering and processing into prepackaged commercial products to be frozen or shipped fresh to markets or wholesalers. Meat-type chickens currently grow to market weight in six to seven weeks, whereas only fifty years ago it took three times as long. This is due to genetic selection and nutritional modifications (and not the use of growth hormones, which are illegal for use in poultry in the US and many other countries). Once a meat consumed only occasionally, the common availability and lower cost has made chicken a common meat product within developed nations. Growing concerns over the cholesterol content of red meat in the 1980s and 1990s further resulted in increased consumption of chicken.

Today, eggs are produced on large egg ranches on which environmental parameters are well controlled. Chickens are exposed to artificial light cycles to stimulate egg production year-round. In addition, it is a common practice to induce molting through careful manipulation of light and the amount of food they receive in order to further increase egg size and production.

On average, a chicken lays one egg a day, but not on every day of the year. This varies with the breed and time of year. In 1900, average egg production was 83 eggs per hen per year. In 2000, it was well over 300. In the United States, laying hens are butchered after their second egg laying season. In Europe, they are generally butchered after a single season. The laying period begins when the hen is about 18–20 weeks old (depending on breed and season). Males of the egg-type breeds have little commercial value at any age, and all those not used for breeding (roughly fifty percent of all egg-type chickens) are killed soon after hatching. The old hens also have little commercial value. Thus, the main sources of poultry meat 100 years ago (spring chickens and stewing hens) have both been entirely supplanted by meat-type broiler chickens.

Transmission of highly pathogenic H5N1 from domestic poultry back to migratory waterfowl in western China has increased the geographic spread. The spread of H5N1 and its likely reintroduction to domestic poultry increase the need for good agricultural vaccines. In fact, the root cause of the continuing H5N1 pandemic threat may be the way the pathogenicity of H5N1 viruses is masked by cocirculating influenza viruses or bad agricultural vaccines.

Webster explains:

If you use a good vaccine you can prevent the transmission within poultry and to humans. But if they have been using vaccines now [in China] for several years, why is there so much bird flu? There is bad vaccine that stops the disease in the bird but the bird goes on pooping out virus and maintaining it and changing it. And I think this is what is going on in China. It has to be. Either there is not enough vaccine being used or there is substandard vaccine being used. Probably both. It's not just China. We can't blame China for substandard vaccines. I think there are substandard vaccines for influenza in poultry all over the world.

In response to the same concerns, Reuters reports Hong Kong infectious disease expert Lo Wing-lok saying that "The issue of vaccines has to take top priority", and Julie Hall, in charge of the WHO's outbreak response in China, saying that China's vaccinations might be "masking" the virus. The BBC reported that Wendy Barclay, a virologist at theUniversity of Reading, UK, said:

The Chinese have made a vaccine based on reverse genetics made with H5N1 antigens, and they have been using it. There has been a lot of criticism of what they have done, because they have protected their chickens against death from this virus but the chickens still get infected; and then you get drift – the virus mutates in response to the antibodies – and now we have a situation where we have five or six "flavours" of H5N1 out there.

Keeping wild birds away from domestic birds is known to be key in the fight against H5N1. Caging (no free range poultry) is one way. Providing wild birds with restored wetlands so they naturally choose nonlivestock areas is another way that helps accomplish this. Political forces are increasingly demanding the selection of one, the other, or both based on nonscientific reasons.

Cattle

Cattle, colloquially referred to as cows, are domesticated ungulates, a member of the family Bovidae, in the subfamily Bovinae, and descended from the aurochs (*Bos primigenius*). They are raised as livestock for meat (called beef and veal), dairy products (milk), leather and as draught animals (pulling carts, plows and the like). In some countries, such as India, they are honored in religious ceremonies and revered. It is estimated that there are 1.4 billion head of cattle in the world today.

Cattle are often raised by allowing herds to graze on the grasses of large tracts of rangeland called ranches. Raising cattle in this manner allows the productive use of land that might be unsuitable for growing crops. The most common interactions with cattle involve daily feeding, cleaning and milking. Many routine husbandry practices involve ear tagging, dehorning, loading, medical operations, vaccinations and hoof care, as well as training for agricultural shows and preparations. There are also some cultural differences in working with cattle - the cattle husbandry of Fulani men rests on behavioural techniques, whereas in Europe cattle are controlled primarily by physical means like fences.

Once cattle obtain an entry-level weight, about 650 pounds (290 kg), they are transferred from the range to a feedlot to be fed a specialized animal feed which consists of corn byproducts (derived from ethanol production), barley, and other grains as well as alfalfa and cottonseed meal. The feed also contains premixes composed of microingredients such as vitamins, minerals, chemical preservatives, antibiotics, fermentation products, and other essential ingredients that are purchased from premix companies, usually in sacked form, for blending into commercial rations. Because of the availability of these products, a farmer using their own grain can formulate their own rations and be assured the animals are getting the recommended levels of minerals and vitamins.

Breeders can utilise cattle husbandry to reduce M. bovis infection susceptibility by selective breeding and maintaining herd health to avoid concurrent disease. Cattle are farmed for beef, veal, dairy, leather and they are sometimes used simply to maintain grassland for wildlife - for example, in Epping Forest, England. They are often used in some of the most wild places for livestock. Depending on the breed, cattle can survive on hill grazing, heaths, marshes, moors and semi desert. Modern cows are more commercial than older breeds and having become more specialised are less versatile. For this reason many smaller farmers still favour old breeds, such as the dairy breed of cattle Jersey. There are many potential impacts on human health due to the modern cattle industrial agriculture system. There are concerns surrounding the antibiotics and growth hormones used, increased E. Coli contamination, higher saturated fat contents in the meat because of the feed, and also environmental concerns.

As of 2010, in the U.S. 766,350 producers participate in raising beef. The beef industry is segmented with the bulk of the producers participating in raising beef calves. Beef calves are generally raised in small herds, with over 90% of the herds having less than 100 head of cattle. Fewer producers participate in the finishing phase which often occurs in a feedlot, but nonetheless there are 82,170 feedlots in the United States.

HOGS

According to Denis Avery of the Hudson Institute, Asia increased its consumption of pork by 18 million tons in the 1990s. As of 1997, the world had a stock of 900 million pigs, which Avery predicts will rise to 2.5 billion pigs by 2050. He told the College of Natural Resources at the University of California, Berkeley that three billion pigs will thereafter be needed annually to meet demand.

AQUACULTURE

Aquaculture is the cultivation of the natural produce of water (fish, shellfish, algae and other aquatic organisms). The term is distinguished from fishing by the idea of active human effort in maintaining or increasing the number of organisms involved, as opposed to simply taking them from the wild. Subsets of aquaculture include Mariculture(aquaculture in the ocean); Algaculture (the production of kelp/seaweed and other algae); Fish farming (the raising of catfish, tilapia and milkfish in freshwater and brackish ponds or salmon in marine ponds); and the growing of cultured pearls. Extensive aquaculture is based on local photosynthetical production while intensive aquaculture is based on fish fed with an external food supply.

Aquaculture has been used since ancient times and can be found in many cultures. Aquaculture was used in China c. 2500 BC. When the waters lowered after river floods, some fishes, namely carp, were held in artificial lakes. Their brood were later fed using nymphs and silkworm feces, while the fish themselves were eaten as a source of protein. The Hawaiian people practiced aquaculture by constructing fish ponds. A remarkable example from ancient Hawaii is the construction of a fish pond, dating from at least 1,000 years ago, at Alekoko. The Japanese practiced cultivation of seaweed by providing bamboo poles and, later, nets and oyster shells to serve as anchoring surfaces for spores. The Romans often bred fish in ponds.

The practice of aquaculture gained prevalence in Europe during the Middle Ages, since fish were scarce and thus expensive. However, improvements in transportation during the 19th century made fish easily available and inexpensive, even in inland areas, causing a decline in the practice. The first North American fish hatchery was constructed onDildo Island, Newfoundland Canada in 1889, it was the largest and most advanced in the world.

Americans were rarely involved in aquaculture until the late 20th century, but California residents harvested wild kelp and made legal efforts to manage the supply starting c. 1900, later even producing it as a wartime resource.

In contrast to agriculture, the rise of aquaculture is a contemporary phenomenon. According to professor Carlos M. Duarte About 430 (97%) of the aquatic species presently in culture have been domesticated since the start of the 20th century, and an estimated 106 aquatic species have been domesticated over the past decade. The domestication of an aquatic species typically involves about a decade of scientific research. Current success in the domestication of aquatic species results from the 20th century rise of knowledge on the basic biology of aquatic species and the lessons learned from past success and failure. The stagnation in the world's fisheries andoverexploitation of 20 to 30% of marine fish species have provided additional impetus to domesticate marine species, just as overexploitation of land animals provided the impetus for the early domestication of land species.

In the 1960s, the price of fish began to climb, as wild fish capture rates peaked and the human population continued to rise. Today, commercial aquaculture exists on an unprecedented, huge scale. In the 1980s, open-netcage salmon farming also expanded; this particular type of aquaculture technology remains a minor part of the production of farmed finfish worldwide, but possible negative impacts on wild stocks, which have come into question since the late 1990s, have caused it to become a major cause of controversy.

In 2003, the total world production of fisheries product was 132.2 million tonnes of which aquaculture contributed 41.9 million tonnes or about 31% of the total world production. The growth rate of worldwide aquaculture is very rapid (greater than 10% per year for most species) while the contribution to the total from wild fisheries has been essentially flat for the last decade.

In the US, approximately 90% of all shrimp consumed are farmed and imported. In recent years salmon aquaculture has become a major export in southern Chile, especially in Puerto Montt and Quellón, Chile's fastest-growing city.

Farmed fish are kept in concentrations never seen in the wild, e.g. 50,000 fish in a 2-acre (8,100 m^2) area, with each fish occupying less room than the average bathtub. This can cause several forms of pollution. Packed tightly, fish rub against each other and the sides of their cages, damaging their fins and tails and becoming sickened with various diseases and infections.

Some species of sea lice have been noted to target farmed coho and farmed Atlantic salmon specifically. Such parasites may have an effect on nearby wild fish. For these reasons, aquaculture operators frequently need to use strong drugs to keep the fish alive (but many fish still die prematurely at rates of up to 30%) and these drugs inevitably enter the environment.

The lice and pathogen problems of the 1990s facilitated the development of current treatment methods for sea lice and pathogens. These developments reduced the stress from parasite/pathogen problems. However, being in an ocean environment, the transfer of disease organisms from the wild fish to the aquaculture fish is an ever-present risk factor.

The very large number of fish kept long-term in a single location produces a significant amount of condensed feces, often contaminated with drugs, which again affect local waterways. However, these effects appear to be local to the actual fish farm site and may be minimal to non-measurable in high current sites.

Integrated Multi-trophic Aquaculture

Integrated Multi-Trophic Aquaculture (IMTA) is a practice in which the by-products (wastes) from one species are recycled to become inputs (fertilizers, food) for another. Fedaquaculture (e.g. fish, shrimp) is combined with inorganic extractive (e.g. seaweed) and organic extractive (e.g. shellfish) aquaculture to create balanced systems for environmental sustainability (biomitigation), economic stability (product diversification and risk reduction) and social acceptability (better management practices).

"Multi-Trophic" refers to the incorporation of species from different trophic or nutritional levels in the same system. This is one potential distinction from the age-old practice of aquatic polyculture, which could simply be the co-culture of different fish species from the same trophic level. In this case, these organisms may all share the same biological and chemical processes, with few synergistic benefits, which could potentially lead to significant shifts in the ecosystem. Some traditional polyculture systems may, in fact, incorporate a greater diversity of species, occupying several niches, as extensive cultures (low intensity, low management) within the same pond. The "Integrated" in IMTA refers to the more intensive cultivation of the different species in proximity of each other, connected by nutrient and energy transfer through water, but not necessarily right at the same location.

Ideally, the biological and chemical processes in an IMTA system should balance. This is achieved through the appropriate selection and proportions of different species providing different ecosystem functions. The co-cultured species should be more than just biofilters; they should also be harvestable crops of commercial value. A working IMTA system should result in greater production for the overall system, based on mutual benefits to the co-cultured species and improved ecosystem health, even if the individual production of some of the species is lower compared to what could be reached in monoculture practices over a short term period.

Sometimes the more general term "Integrated Aquaculture" is used to describe the integration of monocultures through water transfer between organisms. For all intents and purposes however, the terms "IMTA" and "integrated aquaculture" differ primarily in their degree of descriptiveness. These terms are sometimes interchanged. Aquaponics, fractionated aquaculture, IAAS (integrated agriculture-aquaculture systems), IPUAS (integrated peri-urban-aquaculture systems), and IFAS (integrated fisheries-aquaculture systems) may also be considered variations of the IMTA concept.

Shrimp

A shrimp farm is an aquaculture business for the cultivation of marine shrimp or prawns for human consumption. Commercial shrimp farming began in the 1970s, and production grew steeply, particularly to match the market demands of the USA, Japan and Western Europe. The total global production of farmed shrimp reached more than 1.6 milliontonnes in 2003, representing a value of nearly 9 Billion US$. About 75% of farmed shrimp is produced in Asia, in particular in China and Thailand. The other 25% is produced mainly in Latin America, where Brazil is the largest producer. The largest exporting nation is Thailand.

Shrimp farming has moved from China to Southeast Asia into a meat packing industry. Technological advances have led to growing shrimp at ever higher densities, andbroodstock is shipped worldwide. Virtually all farmed shrimp are penaeids (i.e., of the family Penaeidae), and just two species of shrimp—the *Penaeus vannamei* (Pacific white shrimp) and the *Penaeus monodon* (giant tiger prawn)—account for roughly 80% of all farmed shrimp. These industrial monocultures are very susceptible to diseases, which have caused several regional wipe-outs of farm shrimp populations. Increasing ecological problems, repeated disease outbreaks, and pressure and criticism from both NGOs and consumer countries led to changes in the industry in the late 1990s and generally stronger regulation by governments.

REGULATION

In various jurisdictions, intensive animal production of some kinds is subject to regulation for environmental protection. In the United States, a CAFO (Concentrated Animal Feeding Operation) that discharges or proposes to discharge waste requires a permit and implementation of a plan for management of manure nutrients, contaminants, wastewater, etc., as applicable, to meet requirements pursuant to the federal Clean Water Act. Some data on regulatory compliance and enforcement are available. In 2000, the US Environmental Protection Agency published 5-year and 1-year data on environmental performance of 32 industries, with data for the livestock industry being derived mostly from inspections of CAFOs. The data pertain to inspections and enforcement mostly under the Clean Water Act, but also under the Clean Air Act and Resource Conservation and Recovery Act. Of the 32 industries, livestock production was among the top seven for environmental performance over the 5-year period, and was one of the top two in the final year of that period, where good environmental performance is indicated by a low ratio of enforcement orders to inspections. The five-year and final-year ratios of enforcement/inspections for the livestock industry were 0.05 and 0.01, respectively. Also in the final year, the livestock industry was one of the two leaders among the 32 industries in terms of having the lowest percentage of facilities with violations. In Canada, intensive livestock operations are subject to provincial regulation, with definitions of regulated entities varying among provinces. Examples include Intensive Livestock Operations (Saskatchewan), Confined Feeding Operations (Alberta), Feedlots (British Columbia), High-density Permanent Outdoor Confinement Areas (Ontario) and Feedlots or Parcs d'Engraissement (Manitoba). In Canada, intensive animal production, like other agricultural sectors, is also subject to various other federal and provincial requirements.

In the United States, farmed animals are excluded by half of all state animal cruelty laws including the federal Animal Welfare Act. The 28 hour law, enacted in 1873 and amended in 1994 states that when animals are being transported for slaughter, the vehicle must stop every 28 hours and the animals must be let out for exercise, food, and water. The United States Department of Agriculture claims that the law does not apply to birds. The Humane Methods of Livestock Slaughter Act is similarly limited. Originally passed in 1958, the Act requires that livestock be stunned into unconsciousness prior to slaughter. This Act also excludes birds, who make up more than 90 percent of the animals slaughtered for food, as well as rabbits and fish. Individual states all have their own animal cruelty statutes; however many states have a provision to exempt standard agricultural practices.

In the United States there is a growing movement to mitigate the worst abuses by regulating factory farming. In Ohio animal welfare organizations reached a negotiated settlement with farm organizations while in California Proposition 2, Standards for Confining Farm Animals, an initiated law was approved by voters in 2008. Regulations have been enacted in other states and plans are underway for referendum and lobbying campaigns in other states.

An action plan has been proposed by the USDA in February 2009, called the Utilization of Manure and Other Agricultural and Industrial Byproducts. This program's goal is to protect the environment and human and animal health by using manure in a safe and effective manner. In order for this to happen, several actions need to be taken and these four components include: • Improving the Usability of Manure Nutrients through More Effective Animal Nutrition and Management • Maximizing the Value of Manure through Improved Collection, Storage, and Treatment Options • Utilizing Manure in Integrated Farming Systems to Improve Profitability and Protect Soil, Water, and Air Quality • Using Manure and Other Agricultural Byproducts as a Renewable Energy Source

In 2012 Australia's largest supermarket chain, Coles, announced that as of January 1, 2013, they will stop selling company branded pork and eggs from animals kept in factory farms. The nation's other dominant supermarket chain, Woolworths, has already begun phasing out factory farmed animal products. All of Woolworth's house brand eggs are now cage-free, and by mid-2013 all of their pork will come from farmers who operate stall-free farms.

CONTROVERSIES AND CRITICISMS

Advocates of factory farming claim that factory farming has led to the betterment of housing, nutrition, and disease control over the last twenty years, while opponents claim that it harms the environment, creates health risks, and abuses animals.

Animal welfare

Animal welfare impacts of factory farming can include:

- Close confinement systems (cages, crates) or lifetime confinement in indoor sheds

- Discomfort and injuries caused by inappropriate flooring and housing
- Restriction or prevention of normal exercise and most of natural foraging or exploratory behaviour
- Restriction or prevention of natural maternal nesting behaviour
- Lack of daylight or fresh air and poor air quality in animal sheds
- Social stress and injuries caused by overcrowding
- Health problems caused by extreme selective breeding and management for fast growth and high productivity
- Reduced lifetime (longevity) of breeding animals (dairy cows, breeding sows)
- Fast-spreading infections encouraged by crowding and stress in intensive conditions
- Debeaking (beak trimming or shortening) in the poultry and egg industry to avoid pecking in overcrowded quarters
- Forced and over feeding (by inserting tubes into the throats of ducks) in the production of foie gras

Confinement and overcrowding of animals results in a lack of exercise and natural locomotory behavior, which weakens their bones and muscles. An intensive poultry farm provides the optimum conditions for viral mutation and transmission – thousands of birds crowded together in a closed, warm, and dusty environment is highly conducive to the transmission of a contagious disease. Selecting generations of birds for their faster growth rates and higher meat yields has left birds' immune systems less able to cope with infections and there is a high degree of genetic uniformity in the population, making the spread of disease more likely. Further intensification of the industry has been suggested by some as the solution to avian flu, on the rationale that keeping birds indoors will prevent contamination. However, this relies on perfect, fail-safe biosecurity – and such measures are near impossible to implement. Movement between farms by people, materials, and vehicles poses a threat and breaches in biosecurity are possible. Intensive farming may be creating highly virulent avian flu strains. With the frequent flow of goods within and between countries, the potential for disease spread is high.

Confinement and overcrowding of animals' environment presents the risk of contamination of the meat from viruses and bacteria. Feedlot animals reside in crowded conditions and often spend their time standing in their own waste. A dairy farm with 2,500 cows may produce as much waste as a city of 411,000 people, and unlike a city in which human waste ends up at a sewage treatment plant, livestock waste is not treated. As a result, feedlot animals have the potential of exposure to various viruses and bacteria via the manure and urine in their environment. Furthermore, the animals often have residual manure on their bodies when they go to slaughter. Sometimes, even "free-range" animals are mutilated without the use of painkillers.

Depending on the kind of system involved, prevention and control of disease in intensive animal farming commonly use (where appropriate) several of biosecurity, sanitation, surveillance, vaccinations, antibiotics, various measures for control of parasites and other pests, preconditioning, low-stress management, and removal of infected animals. According to a February 2011 FDA report, nearly 29 million pounds of antimicrobials were sold in 2009 for both therapeutic and non-therapeutic use for all farm animal species. The Union of Concerned Scientists estimates that 70% of that amount is for non-therapeutic use.

The large concentration of animals, animal waste, and the potential for dead animals in a small space poses ethical issues. It is recognized that some techniques used to sustain intensive agriculture can be cruel to animals such as mutilation. As awareness of the problems of intensive techniques has grown, there have been some efforts by governments and industry to remove inappropriate techniques.

On some farms, chicks may be debeaked when very young, causing pain and shock. Confining hens and pigs in crates no larger than the animal itself may lead to physical problems such as osteoporosis and joint pain, and psychological problems including boredom, depression, and frustration, as shown by repetitive or self-destructive actions. In the UK, the Farm Animal Welfare Council was set up by the government to act as an independent advisor on animal welfare in 1979 and expresses its policy as five freedoms: from hunger & thirst; from discomfort; from pain, injury or disease; to express normal behavior; from fear and distress.

Fig. Interior of a gestational sow barn

There are differences around the world as to which practices are accepted and there continue to be changes in regulations with animal welfare being a strong driver for increased regulation. For example, the EU is bringing in further regulation to set maximum stocking densities for meat chickens by 2010, where the UK Animal Welfare Minister commented, "The welfare of meat chickens is a major concern to people throughout the European Union. This agreement sends a strong message to the rest of the world that we care about animal welfare." Factory farming is greatly debated throughout Australia, with many people disagreeing with the methods and ways in which the animals in factory farms are treated. Animals are often under stress from being kept in confined spaces and will attack each other. In an effort to prevent injury leading to infection, their beaks, tails and teeth are removed. Many piglets will die of shock after having their teeth and tails removed, because painkilling medicines are not used in these operations. Others say that factory farms are a great way to gain space, with animals such as chickens being kept in spaces smaller than an A4 page·

Less cruel methods of factory farming are still preferable. For example, in the UK, de-beaking of chickens is deprecated, but it is recognized that it is a method of last resort, seen as better than allowing vicious fighting and ultimately cannibalism. Between 60 and 70 percent of six million breeding sows in the U.S. are confined during pregnancy, and for most of their adult lives, in 2 by 7 ft (0.61 by 2.13 m) gestation crates. According to pork producers and many veterinarians, sows will fight if housed in pens. The largest pork producer in the U.S. said in January 2007 that it will phase out gestation crates by 2017. They are being phased out in the European Union, with a ban effective in 2013 after the fourth week of pregnancy. With the evolution of factory farming, there has been a growing awareness of the issues amongst the wider public, not least due to the efforts of animal rights and welfare campaigners. As a result gestation crates, one of the more contentious practices, are the subject of laws in the U.S., Europe and around the world to phase out their use as a result of pressure to adopt less confined practices.

Human health impact

According to the U.S. Centers for Disease Control and Prevention (CDC), farms on which animals are intensively reared can cause adverse health reactions in farm workers. Workers may develop acute and chronic lung disease, musculoskeletal injuries, and may catch infections that transmit from animals to human beings (such as tuberculosis).

Pesticides are used to control organisms which are considered harmful and they save farmers money by preventing product losses to pests. In the US, about a quarter of pesticides used are used in houses, yards, parks, golf courses, and swimming pools and about 70% are used in agriculture. However, pesticides can make their way into consumers' bodies which can cause health problems. One source of this is bioaccumulation in animals raised on factory farms.

"Studies have discovered an increase in respiratory, neurobehavioral, and mental illnesses among the residents of communities next to factory farms."

The CDC writes that chemical, bacterial, and viral compounds from animal waste may travel in the soil and water. Residents near such farms report problems such as unpleasant smell, flies and adverse health effects.

The CDC has identified a number of pollutants associated with the discharge of animal waste into rivers and lakes, and into the air. The use of antibiotics may create antibiotic-resistant pathogens; parasites, bacteria, and viruses may be spread; ammonia, nitrogen, and phosphorus can reduce oxygen in surface waters and contaminate drinking water; pesticides and hormones may cause hormone-related changes in fish; animal feed and feathers may stunt the growth of desirable plants in surface waters and provide nutrients to disease-causing micro-organisms; trace elements such as arsenic and copper, which are harmful to human health, may contaminate surface waters.

Intensive farming may make the evolution and spread of harmful diseases easier. Many communicable animal diseases spread rapidly through densely spaced populations of animals and crowding makes genetic reassortment more likely. However, small family farms are more likely to introduce bird diseases and more frequent association with people into the mix, as happened in the 2009 flu pandemic

In the European Union, growth hormones are banned on the basis that there is no way of determining a safe level. The UK has stated that in the event of the EU raising the ban at some future date, to comply with a precautionary approach, it would only consider the introduction of specific hormones, proven on a case by case basis. In 1998, theEuropean Union banned feeding animals antibiotics that were found to be valuable for human health. Furthermore, in 2006 the European Union banned all drugs for livestock that were used for growth promotion purposes. As a result of these bans, the levels of antibiotic resistance in animal products and within the human population showed a decrease.

The various techniques of factory farming have been associated with a number of European incidents where public health has been threatened or large numbers of animals have had to be slaughtered to deal with disease. Where disease breaks out, it may spread more quickly, not only due to the concentrations of animals, but because modern approaches tend to distribute animals more widely. The international trade in animal products increases the risk of global transmission of virulent diseases such as swine fever, BSE, foot and mouth and bird flu.

In the United States, the use of antibiotics in livestock is still prevalent. The FDA reports that 80 percent of all antibiotics sold in 2009 were administered to livestock animals, and that many of these antibiotics are identical or closely related to drugs used for treating illnesses in humans. Consequently, many of these drugs are losing their effectiveness on humans, and the total healthcare costs associated with drug-resistant bacterial infections in the United States are between $16.6 billion and $26 billion annually.

Methicillin-resistant Staphylococcus aureus (MRSA) has been identified in pigs and humans raising concerns about the role of pigs as reservoirs of

MRSA for human infection. One study found that 20% of pig farmers in the United States and Canada in 2007 harbored MRSA. A second study revealed that 81% of Dutch pig farms had pigs with MRSA and 39% of animals at slaughter carried the bug were all of the infections were resistant to tetracycline and many were resistant to other antimicrobials. A more recent study found that MRSA ST398 isolates were less susceptible to tiamulin, an antimicrobial used in agriculture, than other MRSA or methicillin susceptible *S. aureus*. Cases of MRSA have increased in livestock animals. CC398 is a new clone of MRSA that has emerged in animals and is found in intensively reared production animals (primarily pigs, but also cattle and poultry), where it can be transmitted to humans. Although dangerous to humans, CC398 is often asymptomatic in food-producing animals.

A 2011 nationwide study reported nearly half of the meat and poultry sold in U.S. grocery stores — 47 percent — was contaminated with S. aureus, and more than half of those bacteria — 52 percent — were resistant to at least three classes of antibiotics. Although Staph should be killed with proper cooking, it may still pose a risk to consumers through improper food handling and cross-contamination in the kitchen. The senior author of the study said, "The fact that drug-resistant S. aureus was so prevalent, and likely came from the food animals themselves, is troubling, and demands attention to how antibiotics are used in food-animal production today."

In April 2009, lawmakers in the Mexican state of Veracruz accused large-scale hog and poultry operations of being breeding grounds of a pandemic swine flu, although they did not present scientific evidence to support their claim. A swine flu which quickly killed more than 100 infected persons in that area, appears to have begun in the vicinity of aSmithfield subsidiary pig CAFO (concentrated animal feeding operation).

Environmental impact

Concentrating large numbers of animals in factory farms is a major contribution to global environmental degradation, through the need to grow feed (often by intensive methods using excessive fertiliser and pesticides), pollution of water, soil and air by agrochemicals and manure waste, and use of limited resources (water, energy).

Livestock production is also particularly water-intensive in indoor, intensive systems. Eight percent of global human water use goes towards animal production, including water used to irrigate feed crops.

Industrial production of pigs and poultry is an important source of GHG emissions and is predicted to become more so. On intensive pig farms, the animals are generally kept on concrete with slats or grates for the manure to drain through. The manure is usually stored in slurry form (slurry is a liquid mixture of urine and feces). During storage on farm, slurry emits methane and when manure is spread on fields it emits nitrous oxide and causes nitrogen pollution of land and water. Poultry manure from factory farms emits high levels of nitrous oxide and ammonia.

Large quantities and concentrations of waste are produced. Air quality and groundwater are at risk when animal waste is improperly recycled.

Environmental impacts of factory farming can include:

- Deforestation for animal feed production
- Unsustainable pressure on land for production of high-protein/high-energy animal feed
- Pesticide, herbicide and fertilizer manufacture and use for feed production
- Unsustainable use of water for feed-crops, including groundwater extraction
- Pollution of soil, water and air by nitrogen and phosphorus from fertiliser used for feed-crops and from manure
- Land degradation (reduced fertility, soil compaction, increased salinity, desertification)
- Loss of biodiversity due to eutrophication, acidification, pesticides and herbicides
- Worldwide reduction of genetic diversity of livestock and loss of traditional breeds
- Species extinctions due to livestock-related habitat destruction (especially feed-cropping)

Labor

Small farmers are often absorbed into factory farm operations, acting as contract growers for the industrial facilities. In the case of poultry contract growers, farmers are required to make costly investments in construction of sheds to house the birds, buy required feed and drugs - often settling for slim profit margins, or even losses.

Market concentration

The major concentration of the industry occurs at the slaughter and meat processing phase, with only four companies slaughtering and processing 81 percent of cows, 73 percent of sheep, 57 percent of pigs and 50 percent of chickens. This concentration at the slaughter phase may be in large part due to regulatory barriers that may make it financially difficult for small slaughter plants to be built, maintained or remain in business. Factory farming may be no more beneficial to livestock producers than traditional farming because it appears to contribute to overproduction that drives down prices. Through "forward contracts" and "marketing agreements", meatpackers are able to set the price of livestock long before they are ready for production. These strategies often cause farmers to lose money, as half of all U.S. family farming operations did in 2007. In 1967, there were one million pig farms in America; as of 2002, there were 114,000. Many of the nation's livestock producers would like to market livestock directly to consumers but with limited USDA inspected slaughter facilities, livestock grown locally can not typically be slaughtered and processed locally.

Demonstrations

From 2011 to 2014 each year between 15,000 and 30,000 people gathered under the theme *We are fed up!* in Berlin to protest against industrial livestock production.

INTENSIVE FARMING

Intensive farming or intensive agriculture also known as industrial agriculture is characterized by a low fallow ratio and higher use of inputs such as capital and labour per unit land area. This is in contrast to traditional agriculture in which the inputs per unit land are lower. Intensive animal husbandry involves either large numbers of animals raised on limited land, usually confined animal feeding operations (CAFO) often referred to as factory farms, or managed intensive rotational grazing (MIRG). Both increase the yields of food and fiber per acre as compared to traditional animal husbandry. In a CAFO feed is brought to the animals, which are seldom moved, while in MIRG the animals are repeatedly moved to fresh forage. Intensive crop agriculture is characterised by innovations designed to increase yield. Techniques include planting multiple crops per year, reducing the frequency of fallow years and improving cultivars. It also involves increased use of fertilizers, plant growth regulators, pesticides and mechanization, controlled by increased and more detailed analysis of growing conditions, including weather, soil, water, weeds and pests.

This system is supported by ongoing innovation in agricultural machinery and farming methods, genetic technology, techniques for achieving economies of scale, logistics and data collection and analysis technology. Intensive farms are widespread in developed nations and increasingly prevalent worldwide. Most of the meat, dairy, eggs, fruits and vegetables available in supermarkets are produced by such farms. Smaller intensive farms usually include higher inputs of labor and more often use sustainable intensive methods. The farming practices commonly found on such farms are referred to as appropriate technology. These farms are less widespread in both developed countries and worldwide, but are growing more rapidly. Most of the food available in specialty markets such as farmers markets is produced by these smallholder farms.

HISTORY

Agricultural development in Britain between the 16th century and the mid-19th century saw a massive increase in agricultural productivity and net output.

This in turn supported unprecedented population growth, freeing up a significant percentage of the workforce, and thereby helped enable the Industrial Revolution. Historians cited enclosure, mechanization, four-field crop rotation, and selective breeding as the most important innovations.

Industrial agriculture arose along with the Industrial Revolution. By the early 19th century, agricultural techniques, implements, seed stocks and cultivars had so improved that yield per land unit was many times that seen in the Middle Ages.

The industrialization phase involved a continuing process of mechanization. Horse drawn machinery such as the McCormick reaperrevolutionized harvesting, while inventions such as the cotton gin reduced the cost of processing. During this same period, farmers began to use steam-powered threshers and tractors, although they were expensive and dangerous. In 1892, the first gasoline-powered tractor was successfully developed, and in 1923, the International Harvester Farmall tractor became the first all-purpose tractor, marking an inflection point in the replacement of draft animals with machines. Mechanical harvesters (combines), planters, transplanters and other equipment were then developed, further revolutionizing agriculture. These inventions increased yields and allowed individual farmers to manage increasingly large farms. The identification of nitrogen, potassium, and phosphorus (NPK) as critical factors in plant growth led to the manufacture of synthetic fertilizers, further increasing crop yields. In 1909 the Haber-Bosch method to synthesize ammonium nitrate was first demonstrated. NPK fertilizers stimulated the first concerns about industrial agriculture, due to concerns that they came with serious side effects such as soil compaction, soil erosion and declines in overall soil fertility, along with health concerns about toxic chemicals entering the food supply.

The identification of carbon as a critical factor in plant growth and soil health, particularly in the form of humus, led to so-called *sustainable agriculture,* alternative forms of intensive agriculture that also surpass traditional agriculture, without side effects or health issues. Farmers adopting this approach were initially referred to as *humus farmers*, later as *organic farmers*. The discovery of vitamins and their role in nutrition, in the first two decades of the 20th century, led to vitamin supplements, which in the 1920s allowed some livestock to be raised indoors, reducing their exposure to adverse natural elements. Chemicals developed for use in World War II gave rise to synthetic pesticides.

Following World War II, synthetic fertilizer use increased rapidly, while sustainable intensive farming advanced much more slowly. Most of the resources in developed nations went to improving industrial intensive farming, and very little went to improving organic farming. Thus, particularly in the developed nations, industrial intensive farming grew to become the dominate form of agriculture. The discovery of antibiotics and vaccines facilitated raising livestock in CAFOs by reducing diseases caused by crowding. Developments in logistics and refrigeration as well as processing technology made long-distance distribution feasible.

Between 1700 and 1980, "the total area of cultivated land worldwide increased 466%" and yields increased dramatically, particularly because of selectively bred high-yielding varieties, fertilizers, pesticides, irrigation and machinery. Global agricultural production doubled between 1820 1920; between 1920 and 1950; between 1950 and 1965; and again between 1965 and 1975 to feed a global population that grew from one billion in 1800 to 6.5 billion in 2002. The number of people involved in farming in industrial countries dropped, from 24 percent of the American population to 1.5 percent in 2002. In 1940, each farmworker supplied 11 consumers, whereas in 2002, each worker supplied 90 consumers. The number of farms also decreased and their ownership became more concentrated. In 2000 in the U.S., four companies produce 81 percent of cows, 73 percent of sheep, 57 percent of pigs, and produce 50 percent of chickens, cited as an example of "vertical integration" by the president of the U.S. National Farmers' Union. Between 1967 and 2002 the one million pig farms in America consolidated into 114,000 with 80 million pigs (out of 95 million) produced each year on factory farms, according to the U.S. National Pork Producers Council. According to the Worldwatch Institute, 74 percent of the world's poultry, 43 percent of beef, and 68 percent of eggs are produced this way. Concerns over the sustainability of industrial agriculture, which has become associated with decreased soil quality and over the environmental effects of fertilizers and pesticides have not subsided. Alternatives such as Integrated pest management (IPM), have had little impact because policies encourage the use of pesticides and IPM is knowledge-intensive. These concerns sustained the organic movement and caused a resurgence in sustainable intensive farming and funding for the development of appropriate technology.

Famines continued throughout the 20th century. Through the effects of climactic events, government policy, war and crop failure, millions of people died in each of at least ten famines between the 1920s and the 1990s.

TECHNIQUES AND TECHNOLOGIES

Livestock

Confined animal feeding operations

Intensive livestock farming, also called "factory farming" is a term referring to the process of raising livestock in confinement at high stocking density. "Concentrated animal feeding operations" (CAFO) or "intensive livestock operations", can hold large numbers (some up to hundreds of thousands) of cows, hogs, turkeys or chickens, often indoors. The essence of such farms is the concentration of livestock in a given space. The aim is to provide maximum output at the lowest possible cost and with the greatest level of food safety. The term is often used pejoratively. However, CAFOs have dramatically increased the production of food from animal husbandry worldwide, both in terms of total food produced and efficiency. Food and water is delivered to the animals, and therapeutic use of antimicrobial agents, vitamin supplements and growth hormones are often employed. Growth hormones are not used on chickens nor on any animal in the European Union. Undesirable behaviours often related to the stress of confinement led to a search for docile breeds (e.g., with natural dominance behaviours bred out), physical restraints to stop interaction, such as individual cages for chickens, or physically modification such as the de-beaking of chickens to reduce the harm of fighting.

The CAFO designation resulted from the 1972 US Federal Clean Water Act, which was enacted to protect and restore lakes and rivers to a "fishable, swimmable" quality. TheUnited States Environmental Protection Agency (EPA) identified certain animal feeding operations, along with many other types of industry, as "point source" groundwaterpolluters. These operations were subjected to regulation. In 17 states in the U.S., isolated cases of groundwater contamination were linked to CAFOs. For example, the ten million hogs in North Carolina generate 19 million tons of waste per year. The U.S. federal government acknowledges the waste disposal issue and requires that animal waste be stored in lagoons. These lagoons can be as large as 7.5 acres (30,000 m^2). Lagoons not protected with an impermeable liner can leak into groundwater under some conditions, as can runoff from manure used as fertilizer. A lagoon that burst in 1995 released 25 million gallons of nitrous sludge in North Carolina's New River. The spill allegedly killed eight to ten million fish.

The large concentration of animals, animal waste and dead animals in a small space poses ethical issues to some consumers. Animal rights and animal welfare activists have charged that intensive animal rearing is cruel to animals.

Other concerns include persistent noxious odor, the effects on human health and the role of antibiotics use in the rise of resistant infectious bacteria. According to the U.S. Centers for Disease Control and Prevention (CDC), farms on which animals are intensively reared can cause adverse health reactions in farm workers. Workers may develop acute and/or chronic lung disease, musculoskeletal injuries and may catch (zoonotic) infections from the animals.

Managed intensive rotational grazing

Managed Intensive Rotational Grazing (MIRG), also known as cell grazing, mob grazing and holistic managed planned grazing, is a variety of forage use in which herds/flocks are regularly and systematically moved to fresh, rested grazing areas to maximize the quality and quantity of forage growth. MIRG can be used with cattle, sheep, goats, pigs, chickens, turkeys, ducks and other animals. The herds graze one portion of pasture, or a paddock, while allowing the others to recover. Resting grazed lands allows the vegetation to renew energy reserves, rebuild shoot systems, and deepen root systems, resulting in long-term maximum biomass production. MIRG is especially effective because grazers thrive on the more tender younger plant stems. MIRG also leave parasites behind to die off minimizing or eliminating the need for de-wormers. Pasture systems alone can allow grazers to meet their energy requirements, and with the increased productivity of MIRG systems, the animals obtain the majority of their nutritional needs, in some cases all, without the supplemental feed sources that are required in continuous grazing systems or CAFOs.

Crops

The Green Revolution transformed farming in many developing countries. It spread technologies that had already existed, but had not been widely used outside of industrialized nations. These technologies included "miracle seeds", pesticides, irrigation and synthetic nitrogen fertilizer.

Seeds

In the 1970s scientists created strains of maize, wheat, and rice that are generally referred to as high-yielding varieties (HYV). HYVs have an increased nitrogen-absorbing potential compared to other varieties. Since cereals that absorbed extra nitrogen would typically lodge (fall over) before harvest, semi-dwarfing genes were bred into their genomes. Norin 10 wheat, a variety developed by Orville Vogel from Japanese dwarf wheat varieties, was instrumental in developing wheat cultivars.

IR8, the first widely implemented HYV rice to be developed by the International Rice Research Institute, was created through a cross between an Indonesian variety named "Peta" and a Chinese variety named "Dee Geo Woo Gen." With the availability of molecular genetics in Arabidopsis and rice the mutant genes responsible (*reduced height (rht)*, *gibberellin insensitive (gai1)* and *slender rice (slr1)*) have been cloned and identified as cellular signalling components of gibberellic acid, a phytohormone involved in regulating stem growth via its effect on cell division. Photosynthetic investment in the stem is reduced dramatically as the shorter plants are inherently more mechanically stable. Nutrients become redirected to grain production, amplifying in particular the yield effect of chemical fertilisers. HYVs significantly outperform traditional varieties in the presence of adequate irrigation, pesticides and fertilizers. In the absence of these inputs, traditional varieties may outperform HYVs. They were developed as F1 hybrids, meaning seeds need to be purchased every season to obtain maximum benefit, thus increasing costs.

Crop rotation

Crop rotation or crop sequencing is the practice of growing a series of dissimilar types of crops in the same space in sequential seasons for benefits such as avoiding pathogen and pest buildup that occurs when one species is continuously cropped. Crop rotation also seeks to balance the nutrient demands of various crops to avoid soil nutrient depletion. A traditional component of crop rotation is the replenishment of nitrogen through the use of legumes and green manure in sequence with cereals and other crops. Crop rotation can also improve soil structure and fertility by alternating deep-rooted and shallow-rooted plants. One technique is to plant multi-species cover crops between commercial crops. This combines the advantages of intensive farming with continuous cover and polyculture.

Irrigation

Crop irrigation accounts for 70% of the world's fresh water use.

Flood irrigation, the oldest and most common type, is typically unevenly distributed, as parts of a field may receive excess water in order to deliver sufficient quantities to other parts. Overhead irrigation, using center-pivot or lateral-moving sprinklers, gives a much more equal and controlled distribution pattern. Drip irrigation is the most expensive and least-used type, but delivers water to plant roots with minimal losses. Water catchment management measures include recharge pits, which capture rainwater and runoff and use it to recharge groundwater supplies. This helps in the replenishment of groundwater wells and eventually reduces soil erosion. Dammed rivers creating Reservoirs store water for irrigation and other uses over large areas. Smaller areas sometimes use irrigation ponds or groundwater.

Weed control

In agriculture, systematic weed management is usually required, often performed by machines such as cultivators or liquid herbicide sprayers. Herbicides kill specific targets while leaving the crop relatively unharmed. Some of these act by interfering with the growth of the weed and are often based on plant hormones. Weed control through herbicide is made more difficult when the weeds become resistant to the herbicide. Solutions include:

- Cover crops (especially those with allelopathic properties) that out-compete weeds or inhibit their regeneration.
- Multiple herbicides, in combination or in rotation
- Strains genetically engineered for herbicide tolerance
- Locally adapted strains that tolerate or out-compete weeds
- Tilling
- Ground cover such as mulch or plastic
- Manual removal
- Mowing
- Grazing
- Burning

Terracing

In agriculture, a terrace is a leveled section of a hilly cultivated area, designed as a method of soil conservation to slow or prevent the rapid surface runoff of irrigation water. Often such land is formed into multiple terraces, giving a stepped appearance. The human landscapes of rice cultivation in terraces that follow the natural contours of the escarpments like contour ploughing is a classic feature of the island of Bali and the Banaue Rice Terraces in Banaue, Ifugao, Philippines. In Peru, the Inca made use of otherwise unusable slopes by drystone walling to create terraces.

Rice paddies

A paddy field is a flooded parcel of arable land used for growing rice and other semiaquatic crops. Paddy fields are a typical feature ofrice-growing countries of east and southeast Asia including Malaysia, China, Sri Lanka, Myanmar, Thailand, Korea, Japan, Vietnam, Taiwan, Indonesia, India, and thePhilippines. They are also found in other rice-growing regions such as Piedmont (Italy), the Camargue (France) and the Artibonite Valley (Haiti). They can occur naturally alongrivers or marshes, or can be constructed, even on hillsides. They require large water quantities for irrigation, much of it from flooding. It gives an environment favourable to the strain of rice being grown, and is hostile to many species of weeds. As the only draft animal species which is comfortable in wetlands, the water buffalo is in widespread use in Asian rice paddies. Paddy-based rice-farming has been practiced Korea since ancient times. A pit-house at the Daecheon-ni archaeological site yielded carbonized rice grains and radiocarbon dates indicating that rice cultivation may have begun as early as the Middle Jeulmun Pottery Period (c. 3500-2000 BC) in the Korean Peninsula. The earliest rice cultivation there may have used dry-fields instead of paddies.

The earliest Mumun features were usually located in naturally swampy, low-lying narrow gulleys and fed by local streams. Some Mumun paddies in flat areas were made of a series of squares and rectangles separated by bunds approximately 10 cm in height, while terraced paddies consisted of long irregularly shapes that followed natural contours of the land at various levels. Like today's, Mumun period rice farmers used terracing, bunds, canals and small reservoirs. Some paddy-farming techniques of the Middle Mumun (c. 850-550 BC) can be interpreted from the well-preserved wooden tools excavated from archaeological rice paddies at the Majeon-ni Site. However, iron tools for paddy-farming were not introduced until sometime after 200 BC. The spatial scale of individual paddies, and thus entire paddy-fields, increased with the regular use of iron tools in the Three Kingdoms of KoreaPeriod (c. AD 300/400-668).

A recent development in the intensive production of rice is System of Rice Intensification (SRI). Developed in 1983 by the French Jesuit Father Henri de Laulanié inMadagascar, by 2013 the number of smallholder farmers using SRI had grown to between 4 and 5 million.

Aquaculture

Aquaculture is the cultivation of the natural products of water (fish, shellfish, algae, seaweed and other aquatic organisms). Intensive aquaculture takes place on land using tanks, ponds or other controlled systems or in the ocean, using cages.

Sustainable intensive farming

Sustainable intensive farming practises have been developed to slow the deterioration of agricultural land and even regenerate soil health and ecosystem services, while still offering high yields. Most of these developments fall in the category of organic farming, or the integration of organic and conventional agriculture.

"Organic systems and the practices that make them effective are being picked up more and more by conventional agriculture and will become the foundation for future farming systems. They won't be called organic, because they'll still use some chemicals and still use some fertilizers, but they'll function much more like today's organic systems than today's conventional systems."

Dr. Charles Benbrook Executive director US House Agriculture Subcommittee Director Agricultural Board - National Academy Sciences (FMR)

The System of Crop Intensification (SCI) was born out of research primarily at Cornell University and smallholder farms in India on SRI. It uses the SRI concepts and methods for rice and applies them to crops like wheat, sugarcane, finger millet, and others. It can be 100% organic, or integrated with reduced conventional inputs. Holistic management is a systems thinking approach that was originally developed for reversing desertification. Holistic planned grazing is similar to rotational grazing but differs in that it more explicitly provides a framework for adapting to four basic ecosystem processes: the water cycle, the mineral cycle including the carbon cycle, energy flow and community dynamics (the relationship between organisms in an ecosystem) as equal in importance to livestock production and social welfare. By intensively managing the behavior and movement of livestock, holistic planned grazing simultaneously increases stocking rates and restores grazing land.

Pasture cropping plants grain crops directly into grassland without first applying herbicides. The perennial grasses form a living mulch understory to the grain crop, eliminating the need to plant cover crops after harvest. The pasture is intensively grazed both before and after grain production using holistic planned grazing. This intensive system yields equivalent farmer profits (partly from increased livestock forage) while building new topsoil and sequestering up to 33 tons of CO2/ha/year. The Twelve Aprils grazing program for dairy production, developed in partnership with USDA-SARE, is similar to pasture cropping, but the crops planted into the perennial pasture are forage crops for dairy herds. This system improves milk production and is more sustainable than confinement dairy production. Integrated Multi-Trophic Aquaculture (IMTA) is an example of a holistic approach. IMTA is a practice in which the by-products (wastes) from one species are recycled to become inputs (fertilizers, food) for another. Fed aquaculture (e.g. fish, shrimp) is combined with inorganic extractive (e.g. seaweed) and organic extractive (e.g. shellfish) aquaculture to create balanced systems for environmental sustainability (biomitigation), economic stability (product diversification and risk reduction) and social acceptability (better management practices). Biointensive agriculture focuses on maximizing efficiency such as per unit area, energy input and water input. Agroforestry combines agriculture and orchard/forestry technologies to create more integrated, diverse, productive, profitable, healthy and sustainable land-use systems. Intercropping can increase yields or reduce inputs and thus represents (potentially sustainable) agricultural intensification. However, while total yield per acre is often increased dramatically, yields of any single crop often diminish. There are also challenges to farmers relying on farming equipment optimized for monoculture, often resulting in increased labor inputs.

Vertical farming is intensive crop production on a large scale in urban centers in multi-story, artificially-lit structures that uses far less inputs and produces fewer environmental impacts. An integrated farming system is a progressive biologically integrated sustainable agriculture system such as IMTA or Zero waste agriculture whose implementation requires exacting knowledge of the interactions of multiple species and whose benefits include sustainability and increased profitability. Elements of this integration can include:

- Intentionally introducing flowering plants into agricultural ecosystems to increase pollen-and nectar-resources required by natural enemies of insect pests
- Using crop rotation and cover crops to suppress nematodes in potatoes

CHALLENGES

The challenges and issues of industrial agriculture for society, for the industrial agriculture sector, for the individual farm, and for animal rights include the costs and benefits of both current practices and proposed changes to those practices. This is a continuation of thousands of years of invention in feeding ever growing populations. [W]hen hunter-gatherers with growing populations depleted the stocks of game and wild foods across the Near East, they were forced to introduce agriculture. But agriculture brought much longer hours of work and a less rich diet than hunter-gatherers enjoyed. Further population growth among shifting slash-and-burn farmers led to shorter fallow periods, falling yields and soil erosion. Plowing and fertilizers were introduced to deal with these problems - but once again involved longer hours of work and degradation of soil resources(Boserup, The Conditions of Agricultural Growth,

Allen and Unwin, 1965, expanded and updated in Population and Technology, Blackwell, 1980.). While the point of industrial agriculture is to profitably supply the world at the lowest cost, industrial methods have significant side effects. Further, industrial agriculture is not an indivisible whole, but instead is composed of multiple elements, each of which can be modified in response to market conditions, government regulation and further innovation and has its own side-effects. Various interest groups reach different conclusions on the subject.

BENEFITS

Population growth

Very roughly:

- 30,000 years ago hunter-gatherer behavior fed 6 million people
- 3,000 years ago primitive agriculture fed 60 million people
- 300 years ago intensive agriculture fed 600 million people
- Today industrial agriculture attempts to feed 6 billion people

Table. Estimated world population at various dates, in thousands

Year	World	Africa	Asia	Europe	Central & South America	North America*	Oceania
8000BCE	8 000						
1000 BCE	50 000						
500 BCE	100 000						
1 CE	200,000 plus						
1000	310 000						
1750	791 000	106 000	502 000	163 000	16 000	2 000	2 000
1800	978 000	107 000	635 000	203 000	24 000	7 000	2 000
1850	1 262 000	111 000	809 000	276 000	38 000	26 000	2 000
1900	1 650 000	133 000	947 000	408 000	74 000	82 000	6 000
1950	2 518 629	221 214	1 398 488	547 403	167 097	171 616	12 812
1955	2 755 823	246 746	1 541 947	575 184	190 797	186 884	14 265
1960	2 981 659	277 398	1 674 336	601 401	209 303	204 152	15 888
1965	3 334 874	313 744	1 899 424	634 026	250 452	219 570	17 657
1970	3 692 492	357 283	2 143 118	655 855	284 856	231 937	19 443
1975	4 068 109	408 160	2 397 512	675 542	321 906	243 425	21 564
1980	4 434 682	469 618	2 632 335	692 431	361 401	256 068	22 828
1985	4 830 979	541 814	2 887 552	706 009	401 469	269 456	24 678
1990	5 263 593	622 443	3 167 807	721 582	441 525	283 549	26 687
1995	5 674 380	707 462	3 430 052	727 405	481 099	299 438	28 924
2000	6 070 581	795 671	3 679 737	727 986	520 229	315 915	31 043
2005	6 453 628	887 964	3 917 508	724 722	558 281	332 156	32 998**

An example of industrial agriculture providing cheap and plentiful food is the U.S.'s "most successful program of agricultural development of any country in the world". Between 1930 and 2000 U.S. agricultural productivity (output divided by all inputs) rose by an average of about 2 percent annually causing food prices to decrease. "The percentage of U.S. disposable income spent on food prepared at home decreased, from 22 percent as late as 1950 to 7 percent by the end of the century."

LIABILITIES

Environment

Industrial agriculture uses huge amounts of water, energy, and industrial chemicals; increasing pollution in the arable land, usable water and atmosphere. Herbicides,insecticides and fertilizers are accumulating in ground and surface waters.

"Many of the negative effects of industrial agriculture are remote from fields and farms. Nitrogen compounds from the Midwest, for example, travel down the Mississippi to degrade coastal fisheries in the Gulf of Mexico. But other adverse effects are showing up within agricultural production systems — for example, the rapidly developing resistance among pests is rendering our arsenal of herbicides and insecticides increasingly ineffective.". Agrochemicals and monoculture have been implicated in Colony Collapse Disorder, in which the individual members of bee colonies disappear. Agricultural production is highly dependent on bees to pollinate many varieties of fruits and vegetables.

Social

A study done for the US. Office of Technology Assessment conducted by the UC Davis Macrosocial Accounting Project concluded that industrial agriculture is associated with substantial deterioration of human living conditions in nearby rural communities.

INTENSIVE PIG FARMING

Intensive pig farming is a subset of pig farming and of Industrial animal agriculture, all of which are types of animal husbandry, in which domestic pigs are raised up to slaughter weight. These operations are known as AFO or CAFO in the U.S.

In this system of pig production, grower pigs are housed indoors in group-housing or straw-lined sheds, whilst pregnant sows are housed in sow stalls (gestation crates) or pens and give birth in farrowing crates. The use of sow stalls for pregnant sows has resulted in lower birth production costs; however, this practice has led to more significant animal welfare concerns. Many of the world's largest producers of pigs (US, Canada, Denmark, Mexico) use sow stalls, but some nations (e.g., the UK) and some US states (e.g., Florida, Arizona, and California) have banned their use.

INTENSIVE PIGGERIES

Intensive piggeries are generally large warehouse-like buildings or barns. Indoor pig systems allow the pigs' conditions to be monitored, ensuring minimum fatalities and increased productivity. Buildings are ventilated and their temperature regulated. Most domestic pig varieties are susceptible to sunburn and heat stress, and all pigs lack sweat glands and cannot cool themselves. Pigs have a limited tolerance to high temperatures and heat stress can lead to death. Maintaining a more specific temperature within the pig-tolerance range also maximizes growth and growth-to-feed ratio. Indoor piggeries have allowed pig farming to be undertaken in countries or areas with unsuitable climate or soil for outdoor pig raising (e.g., Australia). In an intensive operation, pigs will no longer need access to a wallow (mud), which is their natural cooling mechanism. Intensive piggeries control temperature through ventilation or drip water systems (dropping water to cool the system).

Pigs are naturally omnivorous and are generally fed a combination of grains and protein sources (soybeans, or meat and bone meal). Larger intensive pig farms may be surrounded by farmland where feed-grain crops are grown. Consequently, piggeries are reliant on the grains industry. Pig feed may be bought packaged, in bulk or mixed on-site. The intensive piggery system, where pigs are confined in individual stalls, allows each pig to be allotted a portion of feed. The individual feeding system also facilitates individual medication of pigs through feed. This has more significance to intensive farming methods, as the proximity to other animals enables diseases to spread more rapidly. To prevent disease spreading and encourage growth, drug programs such asvitamins and antibiotics are administered preemptively.

Indoor systems, especially stalls and pens (i.e., 'dry,' not straw-lined systems) allow for the easy collection of waste. In an indoor intensive pig farm, manure can be managed through a lagoon system or other waste-management system. However, waste smell remains a problem which is difficult to manage. Pigs in the wild or on open farmland are naturally clean animals.

The way animals are housed in intensive systems varies. Breeding sows will spend the bulk of their time in sow stalls (also calledgestation crates) during pregnancy. The use of stalls may be preferred as they facilitate feed management and growth control and prevent pig aggression (e.g., tail biting, ear biting, vulva biting, food stealing). Sows are moved to farrowing crates, with litter, from before farrowing until weaning, to ease management of farrowing and reduce piglet loss from sows lying on them. Dry or open time for sows can be spent in indoor pens or outdoor pens or pastures. Houses should be clean and well ventilated but draught-free.

Piglets can be subjected to castration, tail docking to reduce tail biting, teeth clipping (to reduce injuring their mother's nipples), andearmarking and tattooing for litter identification. Treatments are usually made without pain killers. Weak runts may be slain shortly after birth. Injections with a high availability iron solution often are given, as sow's milk is low in iron. The docking due to tail biting is a common practice in intensive rearing facilities as animals in that environment are more prone to increased levels of aggression and instability.

Piglets are weaned and removed from the sows at between two and five weeks old and placed in sheds, nursery barns or directly to growout barns. Grower pigs – which comprise the bulk of the herd – are usually housed in alternative indoor housing, such as batch pens. Group pens generally require higher stockmanship skills. Such pens will usually not contain straw or other material.

Alternatively, a straw-lined shed may house a larger group (i.e., not batched) in age groups. Larger swine operations use slotted floors for waste removal, and deliver bulk feed into feeders in each pen; feed is available ad libitum.

Many countries have introduced laws to regulate treatment of farmed animals. In the US, the federal Humane Slaughter Act requires pigs to be stunned before slaughter, although compliance and enforcement is questioned. There is concern from animal liberation/welfare groups that the laws have not resulted in a prevention of animal suffering and that there are "repeated violations of the Humane Slaughter Act at dozens of slaughterhouses".

Legislation

Since 2003 EU legislation has:

- Required that pigs be given environmental enrichment, specifically they must have "permanent access to a sufficient quantity of material to enable proper investigation and manipulation activities, such as straw, hay, wood, sawdust, mushroom compost, peat or a mixture of such ..."
- Prohibited routine tail docking. Under the legislation tail docking may only be used as a last resort. The law provides that farmers must first take measures to improve the pigs' conditions and, only where these have failed to prevent tail biting, may they tail dock.

In the US:

- Nine states have banned the use of gestation crates, with Rhode Island being the most recent as of July 2012.
- Discharge from CAFOs is regulated by the federal Environmental Protection Agency (EPA). In 2003 the EPA revised the Clean Water Act (CWA) to include permitting requirements and effluent (discharge) limitations for CAFOs. Final AFO/CAFO regulation issued 2008 revised portions of the 2003 EPA regulated under EPA's National Point Discharge Elimination System (NPDES) permitting program.

Dispute regarding farming methods

Intensive piggeries have been increasingly criticized in preference of free range systems. Such systems usually refer not to a group-pen or shedding system, but to outdoor farming systems. Those that support outdoor systems usually do so on the grounds that they are more animal friendly and allow pigs to experience natural activities (e.g., wallowing in mud, relating to young, rooting soil).

Outdoor systems are usually less economically productive due to increased space requirements and higher morbidity, (though, when dealing with the killing of piglets and other groups of swine, the methods are the same.) They also have a range of environmental impacts, such as denitrification of soil and erosion. Outdoor pig farming may also have welfare implications, for example, pigs kept outside may get sunburnt and are more susceptible to heat stress than in indoor systems, where air conditioning or similar can be used. Outdoor pig farming may also increase the incidence of worms and parasites in pigs. Management of these problems depends on local conditions, such as geography, climate, and the availability of skilled staff.

Transition of an indoor production system to an outdoor system may present obstacles. Some breeds of pig commonly used in intensive farming have been selectively bred to suit intensive conditions. Lean pink-pigmented pigs are unsuited for outdoor agriculture, as they suffer sunburn and heat stress. In certain environmental conditions – for example, a temperate climate – outdoor pig farming of these breeds is possible. However, there are many other breeds of pig suited to outdoor rearing, as they have been used in this way for centuries, such as Gloucester Old Spot and Oxford Forest. Following the UK ban of sow stalls, the British Pig Executive indicates that the pig farming industry in the UK has declined. The increase in production costs has led to British pig-products being more expensive than those from other countries, leading to increased imports and the need to position UK pork as a product deserving a price premium.

In the late '90s Grampian Country Foods, then the UK's largest pig producer, pointed out that pigmeat production costs in the UK were 44 p/kg higher than on the continent. Grampian stated that only 2 p/kg of this was due to the ban on stalls; the majority of the extra costs resulted from the then strength of sterling and the fact that at that time meat and bone meal had been banned in the UK but not on the continent. A study by the Meat and Livestock Commission in 1999, the year that the sow stall ban came into force, found that moving from sow stalls to group housing added just 1.6 pence to the cost of producing 1 kg of pigmeat. French and Dutch studies show that even in the higher welfare group housing systems – ones giving more space and straw – a kg of pigmeat costs less than 2 pence more to produce than in sow stalls.

Criticism of intensive piggeries

Fig. Sows are often confined in gestation crates, which usually does not allow the pig to turn around. Confinement farming methods have come under increasing public scrutiny due to animal welfare and environmentalconcerns.

Sow breeding systems

Organized campaigns by animal activists have focused on the use of the sow stalls, such as the 'gestation crate' (sow stall) and 'farrowing crate'. The sow stall has now been banned in the UK, certain US states, and other European countries, although it remains part of pig production in much of the US and European Union.

Only the sows selected for breeding (i.e., pregnant sows) will spend time in a sow stall. In an intensive system, the sow will be placed in a stall prior to service (mating) and will stay there for at least the start of her pregnancy, when the risk of miscarriage is higher. The length of the sow's gestation is 3 months, 3 weeks, and 3 days. In certain cases, sows may spend this time in the crate. However, a variety of farming systems are used and the time in the crate may vary from 4 weeks to the whole pregnancy.

There is also some criticism of 'farrowing crates'. A farrowing crate houses the sow in one section and her piglets in another. It allows the sow to lie down and roll over to feed her piglets, but keeps her piglets in a separate section. This prevents the large sow from sitting on her piglets and killing them, which is quite common where the sow is not separated from the piglets. Sows are also prevented from being able to move other than between standing and lying. Some models of farrowing crates may allow more space than others, and allow greater interaction between sow and young. Well-designed farrowing pens in which the sow has ample space can be just as effective as crates in preventing piglet mortality. Some crates may also be designed with cost-effectiveness or efficiency in mind and therefore be smaller.

Authoritative industry data indicate that moving from sow stalls to group housing added 2 pence to the cost of producing 1 kg. of pigmeat. Many English fattening pigs are kept in barren conditions and are routinely tail docked. Since

2003 EU legislation has required pigs to be given environmental enrichment and has banned routine tail docking. However, 80% of UK pigs are tail docked.

As of 2015, it will be illegal to use sow crates on New Zealand pig farms.

Effects on traditional rural communities

Common criticism of intensive piggeries is that they represent a corporatization of the traditional rural lifestyle. Critics feel the rise of intensive piggeries has largely replaced family farming. Between 1982 and 1987 some 21% of Iowa hog farmers went out of business. By 1992, another 12% had gone out of business. In large part, this is because intensive piggeries are more economical than outdoor systems, pen systems, or the sty. In many pork-producing countries (e.g., United States, Canada,Australia, Denmark) the use of intensive piggeries has led to market rationalization and concentration. *The New York Times* reported that keeping pigs and other animals in "unnaturally overcrowded" environments poses considerable health risks for workers, neighbors, and consumers.

Waste management and public health concerns

Contaminants from animal wastes can enter the environment through pathways such as through leakage of poorly constructed manure lagoons or during major precipitation events resulting in either overflow of lagoons and runoff from recent applications of waste to farm fields, or atmospheric deposition followed by dry or wet fallout. Runoff can leach through permeable soils to vulnerable aquifers that tap ground water sources for human consumption. Runoff of manure can also find its way into surface water such as lakes, streams, and ponds.

Many contaminants are present in livestock wastes, including nutrients, pathogens, veterinary pharmaceuticals and naturally excreted hormones. Improper disposal of animal carcasses and abandoned livestock facilities can also contribute to water quality problems in surrounding areas of CAFOs.

Exposure to waterborne contaminants can result from both recreational use of affected surface water and from ingestion of drinking water derived from either contaminated surface water or ground water. High-Risk populations are generally the very young, the elderly, pregnant women, and immunocompromised individuals. Dermal contact may cause skin, eye, or ear infections. Drinking water exposures to pathogens could occur in vulnerable private wells.

At Varkensproefcentrum Sterksel in the Netherlands, a pig farm has been created that reuses its waste streams. CO^2 and ammonia from the pig manure are reused to grow algae which in turn are used to feed the pigs.

Another method to reduce the effect on the environment is to switch to other breeds of pig. The enviropig for example is a type of pig with the capability to digest plant phosphorus more efficiently than ordinary pigs. Nutrient-rich runoff from CAFO's can also contribute to Algal blooms in rivers, lakes and seas. The 2009 HAB (harmful Algal Bloom) event off the coast of Brittany, France is attributed to runoff from an intensive pig farm.

North Carolina

As of 2010, North Carolina houses approximately ten million hogs, most of which are located in the eastern half of the state in industrialized CAFOs or Confined Animal Feeding Operations. This was not the case twenty years ago. The initial horizontal integration and the vertical integration that arose in this industry resulted in numerous issues, including issues of environmental disparity, loss of work, pollution, animal rights, and overall general public health. The most remarkable example of swine CAFO monopoly is found in the United States, where in 2001, 50 producers had control over 70% of total pork production. In 2001, the biggest CAFO had just over 710,000 sows.

Originally, Murphy Family Farms horizontally integrated the North Carolina system. They laid the groundwork for the industry to be vertically integrated. Today the hog industry in North Carolina is led by Smithfield Foods, which has expanded into both nationwide and international production.

The environmental justice problems in North Carolina's agroindustrialization of swine production seem to stem from the history of the coastal region's economy, which has relied heavily on black and low-income populations to supply the necessary agricultural labor. The industry's shift from family-owned hog farms to factory hogging has contributed to the frequent targeting of these areas.

This swine production and pollution that accompanies factory hogging is concentrated in the parts of North Carolina that have the highest disease rates, the least access to medical care, and the greatest need for positive education and economic development. Since hog production has become consolidated in the coastal region of N.C., the high water tables and low-lying flood plains have increased the risk and impact of hog farm pollution. A swine CAFO is made up of three parts: the hog house, the "lagoon," and the "spray field." Waste disposal techniques used by small-scale traditional hog farms, like using waste as fertilizer for commercially viable crops, were adopted and expanded for use by CAFOs. Lagoons are supposed to be protected with an impermeable liner, but some do not work properly. This can cause environmental damage, as seen in 1995 when a lagoon burst in North Carolina. This lagoon released 25 million gallons of noxious sludge into North Carolina's New River and killed approximately eight to ten million fish.

The toxins emitted by the swine CAFOs can produce a variety of symptoms and illnesses ranging from respiratory disorders, headaches, and shortness of breath to hydrogen sulfide poisoning, bronchitis, and asthma. The potential for spray field runoff or lagoon leakage puts nearby residents in danger of contaminated drinking water, which can lead to diseases like samonellosis, giardiasis, Chlamydia, meningitis, crytosporidiosis, worms, and influenza.

Denmark

ThePigSite.com stated that IceNews reported that in 2009 the number of pigs that arrived at slaughterhouses with injuries incurred by planks and chains increased. IceNews cited a Copenhagen Post report saying that increasing abuse "may be caused by the new system, introduced in 2006, which rewards" the rushed loading of animals onto vehicles. Over 2008 and 2009, the number of pig abuse cases in Denmark had increased fivefold.

Sometimes sow stalls are used to restrict the movement of sows during pregnancy. This practice is prohibited for pigs exported to the UK. However, the method was found on some Danish farms by British celebrity chef Jamie Oliver in a television programme for the UK's Channel 4 in 2009.

New Zealand

According to Scoop, in 2009 the New Zealand pork industry was "dealt a shameful public relations slap-in-the-face after its former celebrity kingpin, Mike King, outed theirfarming practices as 'brutal,' 'callous' and 'evil'" on a May episode of New Zealand television show *Sunday*. King condemned the "appalling treatment" of factory farmed pigs. King observed conditions inside a New Zealand piggery, and saw a dead female pig inside a sow stall, lame and crippled pigs and others that could barely stand, pigs either extremely depressed or highly distressed, pigs with scars and injuries, and a lack of clean drinking water and food.

4

Reproductive Technology of Animal

Reproduction is the main theme of life. Obviously, without reproductive success an individual (or species) is evolutionarily doomed. All of the organs of the body and their functions may be viewed as structures and processes that enable us to survive in order to reproduce. In this sense each adult individual is an elaborate device for producing eggs or sperm and for bringing them together in the process of fertilization. Sexuality (fertilization) has evolved because (1) it permits a substantial reshuffling of genes (via random segregation and recombination), and (2) it provides a mechanism by which the diploid state can be reproduced after the genetic diversity has been created. This allows an increased genetic variability in populations, and is thus advantageous in evolutionary terms. The concept of the germ line holds that only germ cells are passed from one generation to the next. They do not originate from a germ layer (in the embryonic sense). This idea arose from work in lower forms, but was demonstrated in vertebrates when Bounoure (1934) showed that the vegetal region of fertilized frog eggs contains a material with special staining properties. He was able to trace this cytoplasm, called germ plasm, through cleavage and gastrulation to a few large cells, the primordial germ cells.

These cells were located initially in the prospective endoderm, and migrated to the genital ridge. When those cells were removed or destroyed by irradiation with ultraviolet light, the frogs became sterile. They could produce no germ cells, though they were otherwise normal. If fragments of endoderm from an unirradiated gastrula that contained some of these cells were transplanted to an irradiated host embryo, the resulting frogs were again fertile. The importance of the germ plasm was confirmed when it was shown that ultraviolet irradiation of the vegetal region of the zygote also resulted in sterile frogs. But when vegetal cytoplasm that contained the germ plasm was transferred with a micropipette to an irradiated zygote, it counteracted the effects of the irradiation. The frogs were fertile. Thus, the germ plasm in frog zygotes contains a determinant for germ cell formation that is sensitive to ultraviolet light and can be transferred with the fluid portion of the cytoplasm. There has been considerable controversy concerning whether primordial germ cells, PGCs, arise in the gonads during embryogenesis or are derived from extragonadal origin. Once it was established that germ cell lines in mammals were derived from an extragonadal origin, investigations focused on two questions: Where and when does the germ line arise? Two possibilities exist: (1) germ cells are set aside before embryonic cells become committed to specific pathways, (2) germ cell lines appear de novo from other cells at some point in development.

All the evidence (based on following Green Fluorescent Protein-tagged cells) leads to the conclusion that mammalian germ cells are not morphologically distinct during early development and are induced to form by surrounding cells. There is only fragmentary data for humans, but in mouse the story is coming clear and so far human data are in line with the mouse data. During primitive streak formation, extraembryonic ectoderm cells near the posterior end of the epiblast begin secreting BMPs. This induces some nearby cells to differentiate and express the transmembrane protein *fragilis*. A subset of these cells then begins to express the protein *stella* which is thought to encode an RNA splicing factor (though that is yet unproven). It is this last set of cells that give rise the the PGCs.

It was long thought that PGCs migrate out of the epiblast, into the extraembryonic mesoderm, then back into the embryo via the allantois. That appears not to be true. Based on visualizing fluorescent live cells, the PGCs migrate directly into the embryonic endoderm from the posterior region of the primitive streak. There they find themselves in the developing hindgut. Migration then proceeds from the posterior end of the gut to the genital ridge where they take up residence and complete differentiation. They continue to grow and divide during the migration such that the 10-100 cells that begin the process are ~5000 cell by the time they reach the ridge. Two major questions concerning the migration of PCGs in mammals remain unanswered: 1) What cues PCGs to migrate specifically to the gonadal anlage (*i.e.* the developing gonadal ridge), and 2) what mechanisms do PCGs use to accomplish this migration?

Chemotaxis, with movement being directed by a substance emitted by the developing gonads, is the long suggested cue for PCG migration, but evidence is incomplete. In vitro TGF-b-like proteins secreted by the mouse gonadal ridge are able to induce migration of mouse PGCs. A similar story exists in zebrafish where the known lymphocyte chemoattractant, SDF-1, seems to be used by PGCs. The *in vivo* significance of these observations is still unknown. Alternative mechanisms for directed cell movement include contact guidance, and differential adhesion using molecules likefibronectin. A theory based on contact guidance proposes that motile PCGs move along paths determined by endodermal and mesodermal tissues. The differential adhesion theory proposes the PCGs move to maximize the strength of their adhesive contacts. Most evidence indicates that PCGs move by ameboid motion during the time of active migration. The cell propels itself using an actomyosin-based force generating mechanism with small membrane specializations (filopodia) on the PCG cell membrane acting as attachment points for the cells during migration. The number of PCGs, as defined by alkaline phosphatase staining, changes greatly during the migratory period. After completing migration, PCGs undergo divisional arrest. Regulation of the cell division (both mitotic and meiotic) in PCGs is poorly understood and a major area of research.

The development of the gonads, either testes or ovary, is directly influenced by the localization of the PGCs. And the chromosome complement of a given set of PGCs results in the production of specific gene products that directly effect gonadal differentiation. During embryonic weeks 5-7, in prospective females (chromosomal XX), germ cells localize in the cortex of the indifferent gonad and participate in the formation of an ovary, while the medulla regresses. In males, primordial germ cells (chromosomally XY) migrate deeper to the medullary region of the gonad and are incorporated into the forming testis, and stimulate the primary sex cords to form seminiferous tubules. Reproduction is a marvelous culmination of individual transcendence. Individual organisms come and go, but, to a certain extent, organisms "transcend" time by reproducing offspring. Let's take a look at reproduction in animals.

WHAT IS REPRODUCTION?

In a nutshell, reproduction is the creation of a new individual or individuals from previously existing individuals. In animals, this can occur in two primary ways: through asexual reproduction and through sexual reproduction. Let's look at asexual reproduction.

Asexual Reproduction

In asexual reproduction, one individual produces offspring that are genetically identical to itself. These offspring are produced by mitosis. There are many invertebrates, including sea stars and sea anemones for example, that produce by asexual reproduction. Common forms of asexual reproduction include:

Budding

- In this form of asexual reproduction, an offspring grows out of the body of the parent.
- Hydras exhibit this type of reproduction.

Gemmules (Internal Buds)

- In this form of asexual reproduction, a parent releases a specialized mass of cells that can develop into offspring.
- Sponges exhibit this type of reproduction.

Fragmentation

- In this type of reproduction, the body of the parent breaks into distinct pieces, each of which can produce an offspring.
- Planarians exhibit this type of reproduction.

Regeneration

- In regeneration, if a piece of a parent is detached, it can grow and develop into a completely new individual.
- Echinoderms exhibit this type of reproduction.

Parthenogenesis

This type of reproduction involves the development of an egg that has not been fertilized into an individual.

Animals like most kinds of wasps, bees, and ants that have no sex chromosomes reproduce by this process. Some reptiles and fish are also capable of reproducing in this manner.

Advantages and Disadvantages of Asexual Reproduction

Asexual reproduction can be very advantageous to certain animals. Animals that remain in one particular place and are unable to look for mates would need to reproduce asexually. Another advantage of asexual reproduction is that numerous offspring can be produced without "costing" the parent a great amount of energy or time.

Environments that are stable and experience very little change are the best places for organisms that reproduce asexually. A disadvantage of this type of reproduction is the lack of genetic variation. All of the organisms are genetically identical and therefore share the same weaknesses. If the stable environment changes, the consequences could be deadly to all of the individuals.

Asexual Reproduction in Other Organisms

Animals are not the only organisms that reproduce asexually. Yeasts, plants and bacteria are capable of asexual reproduction as well. Bacterial asexual reproduction most commonly occurs by a kind of cell division called binary fission. Since the cells produced through this type of reproduction are identical, they are all susceptible to the same types of antibiotics.

MANIPULATION OF REPRODUCTION IN ANIMALS

One of the natural laws of life is to be many from one which is possible through reproduction. In animals, sexual reproduction occurs through fertilization of female gametes (ova) by male gametes (sperms) followed by fusion of two nuclei coming from two gametes. Thus sexual reproduction maintains the genetic traits of organism. But offspring develop and born in natural way. To meet the demand of animal products due to gradual increasing propulsion there is need of increase in number of animals too which is possible bio technologically because natural process of increase in number is slow.

In nature female mammals produce only one egg in a month except pigs. Secondly, ruminant females can have only one offspring in a year. Therefore, biotechnology can help to meet increasing demand and need of quality products of animals. In this regard principles of artificial breeding system has been discussed that may be useful further in the area of animal biotechnology.

ARTIFICIAL INSEMINATION

A male animal produces millions of sperms daily. Theoretically, it can inseminate females regularity and produce several offspring. This excess capacity of male has been utilized through developing new technologies for artificial insemination which can be said as the first animal biotechnology.

The most effective factor that has increased the productivity of cattle is the artificial insemination. However, the breeder must replace the nature through artificial insemination if he ensures about the ovulation of female in herd together at a time. In contrast, if a breeder awaits for female to ovulate and then inseminate separately, the importance and economic significance of artificial insemination get reduced. Therefore, through artificial insemination technology increased number of females can be inseminated by a male.

SEMEN AND ITS STORAGE

In addition methods have been developed to produce semen from male by ejaculation. Semen ejaculate is collected and diluted (extended). Sperm motility and there number per millilitre are examined under the microscope. About 0.2 ml bull semen contains about 10 million motile sperms. The diluted sperms may be used fresh within a few days or cryopreserved at -196 Degree Celsius by using liquid nitrogen. The cryopreserved semen can be stored for a long time and easily transported across states or countries. Thus cryopreserved semen of a single male is capable of inseminating thousands of females of a country or other countries. For example, one ejaculate semen of a bull is sufficient to inseminate about 500 cows.

Animals are a source of high protein, food, fibre and leather which men have been using since the time immemorial. In India milk and milk products have become a part of our civilization. Even in any small festival, demand of these two gets increased as several milk-based sweets and food items are prepared. Right from God worship to dining table use of milk and milk products is unavoidable. However, milk production could not keep pace with milk demand. Therefore, to meet milk demand synthetic milk and synthetic khoa are illegally sold in market even after several punishment threats declared by the Government. Due to unaware use of synthetic milk several infants have been reported to lose their lives Production of food from animals is inexpensive and exhausting, and less efficient than plants. The most common farm hoofed animals are cattle, sheep, goats and pigs but unlike plants their management is expensive. Inspite of expenses rendered on nutrition, shelter and management, animals have been our companion since the dawn of civilization. Moreover, there is regional variation on availability of animal protein to human diet due to cost, culture and religion.

In biological sense, despite several similarities animals differ from plants in their reproduction systems such as number of gametes (ova), completion of life cycle, production of products, etc. Except pigs females of all common livestocks produce one egg at about monthly intervals. Therefore, through biotechnological methods discovered in recent years, improvement in reproductive systems of animals, their milk proteins and milk-products, and several novel products which normally lack in animals can be carried out.

Ovulation Control

In many animals it is difficult to find out oestrous (sexual heat) in animals because it persists only for a few hours and occurs mostly at night. After ovulation (which is indicated by oestrous) females are inseminated. But in a herd it would be economical, easy and simplified management if females are inseminated at a times. However, it is possible only when all the female ovulate at a time; in practice it is not possible to get synchronization of oestrous. Moreover, it is possible to bring about ovulation in about 80 per cent of females by using hormones, for example progesteron and/or prostaglandin. These hormones regulate ovulation cycle of female and result in total synchrony of oestrous.

Sperm Sexing

Sperms are produced in the testes of males and ova in female's ovaries. Sperms and ova contain half of chromosomes as compared to somatic cells. An ovum possesses autosomes and one X chromosome. Similarly, a sperm contains autosomes and one Y chromosome. In animals sex is determined genetically *i.e.* by sex chromosomes. X chromosome determines femaleness and Y maleness. All the ova contain X chromosome, whereas a sperm consists of either X or Y chromosomes. One sperm ejaculate contains half X and half Y chromosomes.

In dairy industry demand of females is more than the males. Secondly, females have more desirable characteristics. The livestock industry prefers animals of one sex. Therefore, through artificial insemination technology X and Y chromosomes can be detected and sex of progenies determined accordingly.

There is a fluorescence dye (Hoechst 33342) that stains X and Y chromosomes with different intensities. Thus, these two chromosomes possibly can be separated by using an instrument, fluorescent activated cell sorter (FACS). Sperms are present in the form of suspension. The FACS converts a suspension of sperms into microdroplets. Each droplet consists of a single sperm cell. Individual microdroplet passes through a laser beam. Microdroplets of different intensities are deflected into separate collection tubes as the fluorescence of dye is measured electronically. The sperms separated by using FACS have recently been used and pre-sexed calves have been produced through in vitro fertilization technique.

Moreover, FACS is very expensive and slow. It takes about 24 h to process one semen ejaculate, whereas the sperm cannot remain viable for a long time. Therefore, more refinement in technique is required for its use on a large scale.

Embryo Transfer

In 1890, the first case of developing pregnancy in rabbits through embryo transfer is known in literature. During 1930s the same method was used in goat and sheep. In catties cases of embryo transfer were reported after 1950 (BIOTOL series, 1992). Embryo transfer method cannot be used widely because of its high cost, technical difficulties and limited supply of embryo from superovulated donors. The embryo develops in foster mother (recipient) which simply acts as incubator and does not make any genetic contribution to the offspring.

Secondly, ruminant female carries one pregnancy at a time as only one egg is produced and fertilized with male's sperm. Thus, there is a chance for increasing the number of egg production at a time and transfer of fertilized eggs *i.e.* embryo into uterus of less important foster mothers other than original female in farm animal.

Multiple Ovulation (Superovulation)

The reproductive cycle of ruminant female is such that the ovarian follicle of a non-pregnant female matures and releases single egg at a time. The time of ovulation differs in different animals, for example 21 days in cows and horse, 16 days in sheep and goats, etc. Normally ovulation occurs as a result of circulation of gonadotropic hormone. But by increasing the concentration of hormone the number of egg production gets increased. In well managed domestic catties 8-10 eggs are superovulated; the number may go to 60. However, this depends on health, nutrition, breed of animals and environment in which they live. Price (1991) has reviewed the current practice of multiple ovulation and superovulation. This technique has spread from cattle and sheep to goats, horses and deers.

He has emphasized that:

- General selection for increased litter size is useful only in sheep,
- The gonadotropic hormone induces superovulation in goats, sheeps and cattle but the response varies so much,

- Immunization against ovarian steroid hormones can increase litter size in sheep. Therefore, different molecular forms of follicle stimulation hormones (FSH) should be characterized in the farm animals.

Misra *et al.* have reported multiple ovulation and embryo transfer in Indian buffalo (Bubalus bubalis). After injecting the gonadotrophin the females are induced for superovulation. During follicular phase (second phase of oestrous cycle) about 20 ovarian follicles are induced. Eventually these grow and filled with fluid. The space of follicle which is filled with fluid is called antrum, and such follicles as antral follicles. In normal course, only one follicle develops and releases one egg after maturation. However, before ovulation the follicles laying against surface of ovary looks large sized (8 mm in sheep and pigs, 15 mm in cattle). Therefore, immature oocytes from follicles of donor females are recovered surgically by using laparoscope. The females to be superovulated are frequently injected with prostaglandin F2a (PGF2a) so that synchronized oestrous could develop in them. After 10 days of oestrous they are injected with hormone FSH up to 4 days followed by PGF2a treatment so that oestrous may be maintained. The FSH treatment induces superovulation. The females are artificially inseminated. The eggs are fertilized. After fertilization embryos undergo developmental stages.

Multiple Ovulation with Embryo Transfer (MOET)

After 6-8 days of fertilization (for sheep and goat) the embryos are recovered. During this stage embryos have come to morula or blastula atage and remain in female's oviduct. In cattle the embryos are recovered without surgery by inserting a catheter into oviduct.

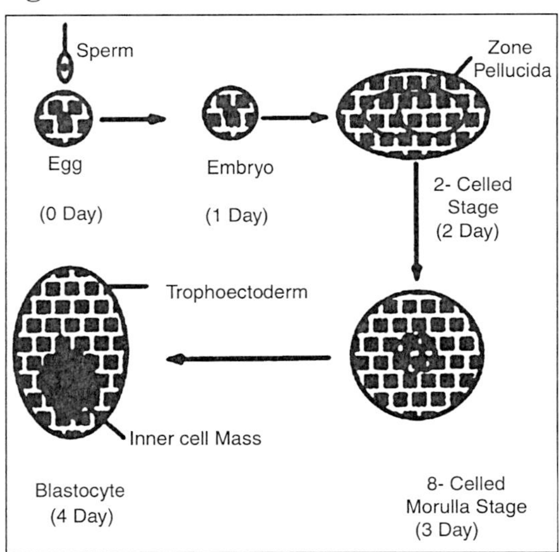

Fig. Developmental Stages of Embryo (Diagrammatic).

A saline solution is drained in oviduct and embryos flushed out through catheter into a storage bottle. The oviduct of sheep is small, therefore, it is exposed surgically and embryos are recovered by syringe adopting the same method for cattle. In storage bottle embryos settle down and solution decanted. Embryos with small volume of saline are poured into a Petri dish. Then they are examined under microscope. The identified embryos are transferred into synchronized recipient females (|as artificial insemination), when they are in 6-8 days of cycle of embryo development, stored and transported or manipulated as desired. In general, one superovulated female results into 5-6 progenies. In cattle 50-60 per cent of pregnancy can be achieved from embryo transfer method.

Embryo Splitting (Demi-Embryo)

Embryo of blastocyst stage is differentiated into two regions, trophoectoderm and inner cell mass (ICM). Trophoectoderm is a single layer of trophoblast cells lining the inner side of zone surrounding the blastoceal (a fluid filled cavity). Zona becomes the foetal membrane bf placenta. The ICM is the mass of cells which develops into foetus. The embryos of blastocyst stage can be split into equal two halves and transferred to females to produce identical twins.

Thus, embryo splitting technology has increased the rate of pregnancy. To carry out embryo splitting, the blastocyst stage embryos are transferred for a few minutes into a cell culture medium consisting of hypertonic sucrose and bovine serum albumin.

The medium of high osmotic strength enters into cell membrane of the embryo (zona pellucida) and cells contract due to exosmosis. The bovine albumin serum attaches to zona pellucida and provides negative charge to the embryo. Then it is transferred into a plastic Petri dish containing standard cell culture medium. The outer membrane of embryo is negatively charged and Petri dish positively. Embryo sticks to the surface of Petri dish due to development of electrostatic interaction of two charges. The Petri dish is kept on stage of an inverted microscope. A micromanipulator equipped with fine surgical blade is used to cut the embryo roughly into two halves with minimal damage to the cells. The hemispherical mass of half embryo, demi-embryo or semi-embryo reforms spheres.

However, it is not necessary for successful implantation of demi-embryo that it should be enclosed in a zona pellucida. The demi-embryos are transferred into oviduct of synchronized recipients as described for normal embryo transfer. At this stage it is very necessary to know the situation for successful embryo implantation that the wall of uterus of synchronized female recipient is waiting for embryo, and embryo is prepared to send chemical signals from the trophoblast.

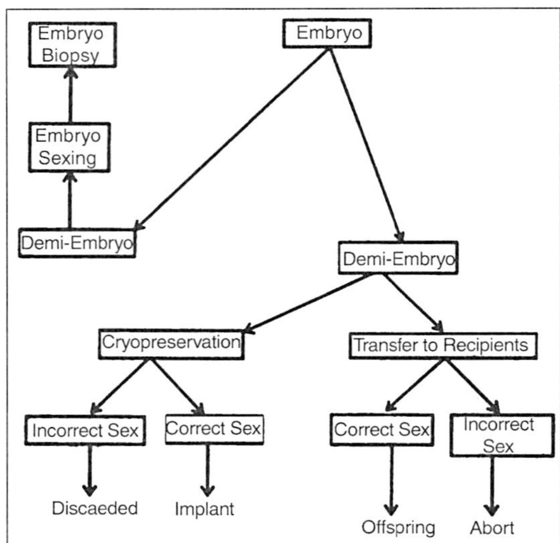

Fig. An Outline of Embryo Splitting.

To increase the rate of pregnancy, the embryo can be cut into equal three or four pieces, and transplanted in synchronized females. But too small trophoblast will not induce the uterus for pregnancy. So far it is unknown how cells of ICM are required for successful pregnancy, but it is clear that increasing in embryo splitting will have less probability of pregnancy.

Embryo biopsy (removal of small number of cells for genetic analysis) should be combined with splitting so that the twins which will be produced should be identical and of known genotype. Biopsy is very necessary in breeding so that sex and genetic diseases could be detected. In case the embryo has any genetic disease, it can be prevented from implantation in recipient females.

Embryo Sexing

Before the implantation of embryo its sex should be detected from the biopsy sample. The principle for sexing is very common. The presence of Y chromosome makes the offsprings male and that of X makes female. The second method is the use of polymerase chain reaction (PCR) machine in sex detection. PCR amplifies DNA sequence of Y chromosome and reaction products can be seen directly. Handy side *et al.* isolated single blastomere from early embryo from a womb, amplified DNA sequences of Y chromosome and carried out embryo sexing before implantation into uterus. It is true that the PCR was first commercially implemented for embryo sexing of livestock. Now the sexed embryos are available commercially but these are very costly.

INFERTILITIES IN HUMANS

There is a large number of reasons for developing infertilities in humans. By using IVF many more infertilities have been treated.

Male Sterility

In a fertile human the total count of sperms should be 15-20 million per ml. When the number of sperms decreases, the person is called oligospermic and the condition as oligospermia. Similarly, when there is very low number of motile sperms, the patient is called azoospermic and the phenomenon as azoospermia. Generally in azoospermic humans the number of non-motile sperms is high. In addition, for IVF, the presence of motile sperms should be around 10-20,000 per ml.

Female Sterility

There are several reasons and types of female infertility; few of them are:

- Non-functional (inaccessible or absent) ovaries (some times ovaries of female are non-functional or inaccessible, therefore, healthy oocytes will not be produced. In such cases oocytes can be procured from donor for IVF),
- Non-functional (or absent) uterus (in this condition the oocytes are obtained from female and subjected to IVF, and embryo transferred to surrogate mother for pregnancy and further development),
- Idiopathic infertility (no definite cause is known for idiopathic infertility. Possibly it may be due to abnormal fertilization or failure in fertilization. Therefore, IVF followed by embryo transfer will cure this infertility),
- Tubal infertility (fallopian tubes receive oocytes from follicles of ovaries for fertilization. Any deformity in fallopian tubes results in tubal infertility. IVF followed by embryo transfer in uterus has replaced the function of fallopian tubes. Therefore, the patients will not be subjected to surgical treatment).

WHO BENEFITS IVF

It is not likely that any patient of any age or any group of the society can go for treatment for IVF. There are several prerequisite conditions applicable for male and female partners.

The male should not be azoospermic; however, if he is azoospermic and desires for a baby, he can procure semen either from sperm bank or elsewhere. Besides, the female partner should also be healthy and fit for IVF because it is she who is to bear the pregnancy.

Therefore, the female must be such that:

- Any surgical procedure should not create any trouble to her,
- Her ovaries should be accessible for oocytes,
- Her uterus should accept and bear pregnancy for the required period of about nine months,
- Her cervical canal should be easily negotiable for embryo transfer.

HOW THE PATIENTS FOR IVF TREATED

First of all the female aspirant for IVF is suggested to keep the records of her menstrual cycle upto six months. This information supports the doctors to take fruitful decision for the possibility of IVF. The female patient is hospitalized to ensure the date for collection of urine for the presence of luteinizing hormone (LH) surge, and to fix date for administrating human choripnic gonadotrophin (hCG). The hCG controls the final stage of formation of follicle and oocytes. The oocytes may be recovered during any menstrual cycles *i.e.* natural cycle, stimulated cycle or controlled cycle. According to the nature of three cycles, the treatment for IVF is recommended.

During natural cycle, spontaneous LH surge is found out at 3-6 h intervals by taking urine or plasma sample. Under the influence of LH surge fully developed follicles of both the ovaries burst and ova collected in fallopian tubes. In natural cycle, only one ova at a time is formed. Secondly, the stimulated cycle is created by administrating clomiphene and/or human menopausal gonadotrophin (hMG). Clomiphene (150 mg) per day is given for the 5th to 9th day of the cycle. However, the duration may be prolonged for 10 minutes but there will be more risk of uterus receptivity and abortion. The hMG alone or along with Clomiphene is also given that does not allow the high rate of abortion.

But several abnormalities are also caused by hMG. Consequently multiple follicles and oocytes, are stimulated instead of one. At this time hCG is also given to female patients so that inhibition of LH surge can be checked. The disadvantages associated with these are the ultrasonographic assessment of the hormonal effect and development of some of abnormal oocytes that result in abnormal pregnancy. Thirdly, hCG is administered at the optimum stage of maturation that arrests the development of follicle. However, it is difficult to predict the optimum time for hCG administration that varies for different preovulatory follicles. Patients need not stay in hospital for a long duration because the doctor can (to his own and patient's convenience) recover oocytes through laparoscopy. His success rate in oocyte recovery, embryo transfer and pregnancies are achieved by the stimulated and controlled cycles.

Indicators for Ovary Stimulation

It is very necessary to monitor the stimulation of ovary so that the oocytes can be collected at right time of development.

There are several parameters which are used as indicators for stimulation of ovary, these include:

- Prediction of LH surge on the basis of menstrual cycle,
- Rise in temperature of body during preovulatory and postovulatory days,
- Changes in cervical mucus as it is secreted five days before LH surge, goes maximum a day before LH surge and falls sharply thereafter,

- Estimation of progesterone,
- Levels of oestrogen in urine, etc.

Oocyte Recovery and Culture

Oocytes are recovered through recently developed equipment called laparoscopy. Using microscopic equipment the follicles can be observed and aspirated with the aspirating apparatus which consists of 23 cm long needle made up of stainless steel (with internal diameter of 1.6 mm and external diameter 2.1 mm) and 52 cm long tubing. The total space between needle and tubing is 1.2 ml. The ovulating follicles are selected for aspiration. It consists of ova of less than 1.5 cm. From such follicles fluid (4-12 ml) is aspirated by using needle and tubing. After aspiration wall of follicle gets collapsed. Fluid is of straw coloured some times it contains blood also. When there appears blood, heparinised culture medium should be added in tube so that blood clotting may be prevented. However, if fluid contains blood, ova could not be easily identified and isolated.

The aspirate is observed under the microscope for the presence of oocytes. If ova are not present in aspirate, the heparinised culture medium should be reflushed into follicles to get oocytes. The oocytes are identified under the microscope and incubated for 5-10 h depending on maturity of oocytes and follicles. The media used for oocyte culture and IVF are modified Ham's F10 medium, modified Whitten's medium, Earl's solution and Whittingham's T6 medium. Any one of these media may be used.

Semen Preparation

The semen is procured either (from husband of female partner or from semen bank or elsewhere. Husband's semen is collected at site when required through masturbation 60-90 minutes prior to insemination. Semen is subjected to centrifugation; sperm pellets are recovered and transferred to culture medium for suspending it again. It is then centrifuged and semen pellets are retransferred to culture medium. The sperms are incubated at 37°C for 30-60 minutes. The most active spermatozoa float at surface. Therefore, samples for IVF should be taken from surface of culture medium containing spermatozoa.

ANIMAL REPRODUCTIVE SYSTEM

Animal reproductive system, any of the organ systems by which animals reproduce. The role of reproduction is to provide for the continued existence of a species; it is the process by which living organisms duplicate themselves. Animals compete with other individuals in the environment to maintain themselves for a period of time sufficient to enable them to produce tissue nonessential to their own survival, but indispensable to the maintenance of the species. The additional tissue, reproductive tissue, usually becomes separated from the individual to form a new, independent organism.

The reproductive systems in metazoans (multicelled animals) from sponges to mammals, exclusive of humans. It focuses on the gonads (sex organs), associated ducts and glands, and adaptations that aid in the union of gametes—*i.e.*, reproductive cells, male or female, that are capable of producing a new individual by union with a gamete of the opposite sex. Brief mention is made of how the organism provides for the development of embryos and of the regulatory role of gonads in vertebrate cycles.

Unlike most other organ systems, the reproductive systems of higher animals have not generally become more complex than those of lower forms. Asexual reproduction (*i.e.*, reproduction not involving the union of gametes), however, occurs only in the invertebrates, in which it is common, occurring in animals as highly evolved as the sea squirts, which are closely related to the vertebrates. Temporary gonads are common among lower animals; in higher animals, however, gonads are permanent organs. Hermaphroditism, in which one individual contains functional reproductive organs of both sexes, is common among lower invertebrates; yet separate sexes occur in such primitive animals as sponges, and hermaphroditism occurs in animals more highly evolved—*e.g.*, the lower fishes. Gonads located on or near the animal surface are common in the lowest invertebrates, but in higher animals they tend to be more deeply situated and often involve intricate duct systems. In echinoderms, which are among the highest invertebrates, the gonads hang directly into the sea and spill their gametes into the water. In protochordates, gametes are released into a stream of respiratory water that passes directly into the sea. Duct systems of the invertebrate flatworms (Platyhelminthes) are relatively complex, and those of specialized arthropods (*e.g.*, insects, spiders, crabs) are more complex than those of any vertebrate. Copulatory organs occur in flatworms, but copulatory organs are not ubiquitous among vertebrates other than reptiles and mammals. The trend towards fewer eggs and increased parental care in higher animals may account for the relative lack of complexity in the reproductive systems of some advanced forms. Whereas trends towards increasing structural complexity have often been reversed during evolution, reproductive behaviour patterns in many phylogenetic (*i.e.*, evolutionary) lines have become more complicated in order to enhance the opportunity for fertilization of eggs and maximum survival of offspring.

A direct relationship exists between behaviour and the functional state of gonads. Reproductive behaviour induced principally but not exclusively by organic substances called hormones promotes the union of sperm (spermatozoa) and eggs, as well as any parental care accorded the young. There are a number of reasons why behaviour must be synchronized with gonadal activity. Chief among these are the following:

Individuals of a species must congregate at the time the gonads contain mature gametes. This often entails migration, and some members of all major vertebrate groups migrate long distances to gather at spawning grounds or rookeries. Individuals with gametes ready to be shed must recognize members of the opposite sex. Recognition is sometimes by external appearance or by chemical substances (pheromones), but sex-linked behaviour is often the only signal.

Geographical territories frequently must be established and aggressively defended. The building of nests, however simple, is essential reproductive behaviour in many species. When fertilization of aquatic forms is external, sperm and eggs must be discharged at approximately the same time into the water, since gametes may be quickly dispersed by currents. Courtship, often involving highly intricate behaviour patterns, serves to release the gametes of both mating individuals simultaneously.

When fertilization is internal, willingness of the female to mate is often essential. Female mammals not in a state of willingness to mate not only will not mate but may injure or even kill an aggressive male. The unwillingness of a female mammal to mate when mature eggs are not present prevents loss of sperm needed to preserve the species. Parental care of fertilized eggs by one parent or the other has evolved in many species. Parental behaviour includes fanning the water or air around the eggs, thereby maintaining appropriate temperature and oxygen levels; secretion of oxygen from a parent's gills; transport of eggs on or in the parental body (including the mouth of some male parents); and brooding, or incubation, of eggs.

Some species extend parental care into the postnatal period, feeding and protecting the offspring. Such behaviour patterns are adaptations for survival and thus are essential; all are induced by the nervous and endocrine systems and are typically cyclical, because gonadal activity is cyclical.

REPRODUCTIVE SYSTEMS OF INVERTEBRATES

GONADS, ASSOCIATED STRUCTURES, AND PRODUCTS

Although asexual reproduction occurs in many invertebrate species, most reproduce sexually. The basic unit of sexual reproduction is a gamete (sperm or egg), produced by specialized tissues or organs called gonads. Sexual reproduction does not necessarily imply copulation or even a union of gametes. As might be expected of such a large and diverse group as the invertebrates, many variations have evolved to ensure survival of species. In many lower invertebrates, gonads are temporary organs; in higher forms, however, they are permanent. Some invertebrates have coexistent female and male gonads; in others the same gonad produces both sperm and eggs. Animals in which both sperm and eggs are produced by the same individual (hermaphroditism) are termed monoecious. In dioecious species, the sexes are separate. Generally, the male gonads ripen first in hermaphroditic animals (protandry); this tends to ensure cross-fertilization. Self-fertilization is normal, however, in many species, and some species undergo sex reversal.

SPONGES, COELENTERATES, FLATWORMS, AND ASCHELMINTHS

Sponges are at a cellular level of organization and thus do not have organs or even well-developed tissues; nevertheless, they produce sperm and eggs and also reproduce asexually. Some species of sponge are monoecious, others are dioecious. Sperm and eggs are formed by aggregations of cells called amoebocytes in the body wall; these are not considered gonads because of their origin and transitory nature.

In hydrozoan coelenterates, temporary gonads are formed by groups of cells in either the epidermis (outer cell layer) or gastrodermis (gut lining), depending on the species; scyphozoan and anthozoan coelenterates generally have gonads in the gastrodermis. The origin and development of gonads in coelenterates, particularly freshwater species, are often associated with the seasons. Freshwater hydrozoans, for example, reproduce asexually until the onset of cold weather, which stimulates them to form testes and ovaries. Colonial hydrozoans asexually produce individuals known as polyps. Polyps, in turn, give rise to free swimming stages (medusae), in which gonads develop. The body organization of sponges and coelenterates is such that most of their cells are in intimate contact with the environment; consequently, gametes are shed into the water, and no ducts are necessary to convey them to the outside.

In contrast to sponges and coelenterates, platyhelminths generally have well-developed organ systems of a permanent nature and, in addition, have evolved secondary reproductive structures to convey sex products. One exception is the acoels, a group of primitive turbellarians; they lack permanent gonads, and germinal cells develop from amoebocytes in much the same manner as in sponges. The majority of flatworms, however, are monoecious, the primary sex organs consisting of one or more ovaries and testes. The tube from the ovary to the outside is called the oviduct; it often has an outpocketing (seminal receptacle) for the storage of sperm received during copulation. In many species the oviduct receives a duct from yolk (vitelline) glands, whose cells nourish the fertilized egg. Beyond the entrance of the duct from the yolk glands the oviduct may be modified to secrete a protective capsule around the egg before it is discharged to the outside. The male organs consist of testes, from which extend numerous tubules (vasa efferentia) that unite to form a sperm duct (vas deferens); the latter becomes an ejaculatory duct through which sperm are released to the outside. The sperm duct may exhibit expanded areas that store sperm (seminal vesicles), and it may be surrounded by prostatic cells that contribute to the seminal fluid. The sperm duct eventually passes through a copulatory organ. The same basic structural pattern, somewhat modified, is found in most higher invertebrates.

Aschelminthes (roundworms) are mostly dioecious; frequently there are external differences between males and females (sexual dimorphism). The males are generally smaller and often have copulatory spicules. Nematodes have relatively simple reproductive organs, a tubular testis or ovary being located at the end of a twisted tube. The portion of the female tract nearest the ovary forms a uterus for temporary storage of fertilized eggs. Some species lay eggs, but others retain the egg in the uterus until the larva hatches. The sperm are released into a cavity called the cloaca. A number of free-living nematodes are capable of sex reversal—if the sex ratio in a given population is not optimal or if environmental conditions are not ideal, the ratio of males to females can be altered. This sometimes results in intersexes; *i.e.*, females with some male characteristics. Hermaphroditism occurs in nematodes, and self-fertilization in such species is common.

ANNELIDS AND MOLLUSKS

Annelids have a well-developed body cavity (coelom), a part of the lining of which gives rise to gonads. In some annelids, gonads occur in several successive body segments. This is true, for example, in polychaetes, most of which are dioecious. Testes and ovaries usually develop, though not invariably, in many body segments; and the sperm and eggs, often in enormous numbers, are stored in the coelom. Fertilization is external. In oligochaetes (all of which are monoecious) on the other hand, the gonads develop in a few specific segments. Sperm are stored in a seminal vesicle and eggs in an egg sac, rather than in the coelom. A portion of the peritoneum, the membrane lining the coelom, becomes a saclike seminal receptacle that stores sperm received from the mate. The earthworm, Lumbricus terrestris, is an example of a specialized annelid reproductive system.

Female organs consist of a pair of ovaries in segment 13; a pair of oviducts that open via a ciliated funnel (*i.e.*, with hairlike structures) into segment 13 but open to the exterior in segment 14; an egg sac near each funnel; and a pair of seminal receptacles in segment 9 and another in segment 10. Male organs consist of two pairs of minute testes in segments 10 and 11, each associated with a ciliated sperm funnel leading to a tiny duct, the vas efferens. The two ducts on each side lead to a vas deferens that opens in segment 15. Testes and funnels are contained within two of three pairs of large seminal vesicles that occupy six body segments. Leeches (Hirudinea), also monoecious, have one pair of ovaries and a segmentally arranged series of testes with duct systems basically similar to those of earthworms.

Although mollusks have a close evolutionary kinship to annelids, they have reduced or lost many structures characteristic of segmented worms. The coelom persists only as three regional cavities: gonadal, nephridial (kidney), and pericardial (heart). In ancestral forms these were interconnected so that gametes from the gonad passed through the pericardial cavity, the nephridial cavity, then to the outside through a nephridial pore. The various groups of mollusks have tended to modify this arrangement, with the result that gonads have their own pore; among amphineurans, for example, the sexes are usually separate, and there is one gonad with an associated pore. Gastropods show considerable variability, but generally one gonad (ovary, testis, or ovotestis—a structure combining the functional gonads of both sexes) is located in the visceral hump and connected to the outside by a remnant of the right kidney. In hermaphroditic forms, one duct carries sperm as well as eggs.

The gonadal ducts of gastropod females often secrete a protective capsule around the fertilized eggs; in males, the terminal portion of the duct is sometimes contained in a copulatory organ. Pelecypods may be either monoecious or dioecious, but the gonads are usually paired. In mussels and oysters, the gonads open through the nephridial pore, but in clams the reproductive system opens independently. The cephalopods are all dioecious. The single testis or ovary releases its products into the pericardial cavity and this, in turn, leads to a gonopore, the external opening. The oviduct of the squid is terminally modified to form a shell gland. The male system is more complex—the gonoduct leads into a seminal vesicle where a complicated torpedo-shaped sperm case (spermatophore) is secreted and contains the sperm. Spermatophores are then stored in a special structure (Needham's sac) until copulation occurs.

A remarkable characteristic of some mollusks is the ability to alter their sex. Some species are clearly dioecious; however, among the monoecious species there is considerable variability in their hermaphroditic condition. In some species, male and female gonads, although in the same individual, are independent functionally and structurally. In others, an ovotestis produces both sperm and eggs.

Oysters display a third condition; young oysters have a tendency towards maleness, but, if water temperature or food availability is altered, some individuals develop into females. Later, a reversal to the male condition may occur. The sexual makeup of an entire oyster population also has a seasonal aspect; in harmony with the group, an individual may undergo several alterations in the course of a year. A similar phenomenon, called consecutive sexuality, occurs in limpets. These gastropods stack themselves in piles, with the younger animals on top. The animals on top are males with well-developed testes and copulatory organs; those in the middle are hermaphroditic; those on the bottom are females, having lost the testes and copulatory organ (penis) by degeneration. A decrease in the number of females in a stack induces males to assume female characteristics, but the transition is retarded when an excess of females is present. The degree of maleness or femaleness is probably controlled in part by environmental and internal factors.

ARTHROPODS

The phylum Arthropoda includes a vast number of organisms of great diversity. Most arthropods are dioecious, but many are hermaphroditic, and some reproduce parthenogenetically (*i.e.*, without fertilization). The primary reproductive organs are much the same as in other higher invertebrates, but the secondary structures are often greatly modified. Such modifications depend on whether fertilization is internal or external, whether the egg or zygote (*i.e.*, the fertilized egg) is retained or immediately released, and whether eggs are provided some means of protection after they have left the body of the female. The mandibulate arthropods (*e.g.*, crustaceans, insects) include more species than any other group and have invaded most habitats, a fact reflected in their reproductive processes.

Crustaceans (*e.g.*, crabs, crayfish, barnacles) are for the most part dioecious. The primary reproductive organs generally consist of paired gonads that open through paired ventral (bottom side) gonopores. Females often have a seminal receptacle (spermatheca) in the form of an outpocketing of the lower part of the female tract or as an invagination (inpocketing) of the body near the gonopore. Males have appendages modified for clasping the female during copulation or for guiding sperm. A number of groups have members that reproduce parthenogenetically. Branchiopods (*e.g.*, water fleas, fairy shrimp) have simple paired gonads. The female gonopore often opens dorsally (on the back side) into a brood chamber; the male gonopore opens near the anus. Males have appendages for clasping females during copulation. Ostracods, or seed shrimp, have paired, tubular gonads. The eggs may be brooded by the female, or they may be released into the water via a gonoduct and gonopore. The terminal portion of the male gonoduct is enclosed in a single or paired penis. Many species reproduce parthenogenetically. Some experts contend that this is the only method employed, even though functional males may be present in the population. Copepods (*e.g.*, Cyclops) have paired ovaries and an unpaired testis. The terminal portion of the oviduct constitutes an ovisac for storage of eggs. The male deposits sperm in a spermatophore that is transferred to the female. Sexual dimorphism is particularly evident among parasitic copepods. Frequently, parasitic females can hardly be recognized as copepods except for the distinctive ovisacs. Males, on the other hand, are free-living and are recognizable as copepods.

The hermaphroditic Cirripedia (*e.g.*, barnacles) are among the exceptions to the generalization that crustaceans are dioecious. It has been suggested that hermaphroditism in barnacles is an adaptation to their sessile, or stationary, existence, but cross-fertilization is more common than self-fertilization. The ovaries lie either in the base or in the stalk of the animal, and the female gonopore is near the base of the first pair of middle appendages (cirri). The testes empty into a seminal vesicle through a series of ducts; from the vesicle extends a long sperm duct within a penis that may be extended to deposit sperm in the mantle cavity of an adjacent barnacle. The terminal portion of the oviduct secretes a substance that forms a kind of ovisac within the mantle cavity, where fertilized eggs undergo early development. Although most barnacles are hermaphrodites, some display a peculiar adaptation in that they contain parasitic dwarf or accessory males. Dwarf males are much smaller than the host barnacle in which they live and are degenerate, except for the testes. In some species they live in the mantle cavity of hermaphroditic forms and produce accessory sperm; in other species only the female organs persist in the host animal, and the accessory male is a necessity.

Amphipods and isopods (*e.g.*, pill bugs, sow bugs), like most crustaceans, are dioecious and have paired gonads. Females of both groups have a ventral brood chamber (marsupium) formed by a series of medially directed (*i.e.*, towards the body midline) plates (oostegites) in the region of the thorax, the region between head and abdomen. Many isopods are parasitic and have developed unusual sex-related activities. Certain species are parasitic on other crustaceans. After a series of molts (*i.e.*, shedding of the body covering) a parasitic larval (immature) isopod attaches to the shell of a crab. If it is the only larva to do so, it increases in size and develops into an adult female. If another larva subsequently attaches, the new arrival becomes a male. It has been demonstrated that the testes of the functioning male larva will change to ovaries if the larva is removed to a new, uninfected host. Thus, the larvae of these species apparently are intersexual and can develop into either sex. This phenomenon, reminiscent of that in mollusks, demonstrates the way in which similar adaptations have evolved in diverse groups of organisms.

The gonads of crabs and lobsters are paired, as are the gonopores. The females of many species have external seminal receptacles on the ventral part of the thorax; those of other species have internal receptacles in the same region. In some species, seminal receptacles are absent, and the male simply attaches a spermatophore to the female. Thus, males either have appendages (gonopods) by which sperm are inserted in the body of the female or produce spermatophores for sperm transfer. The sexual dimorphism of many decapods can be altered by parasitism.

An example of this is the crab that is parasitized by a barnacle. A barnacle infection in male crabs induces the secondary sex characters of the crab to resemble those of a female; however, masculinization does not occur in parasitized females. At each molt a parasitized male crab increasingly resembles a female even though the testes may be completely unaffected. Feminization results from a hormonal alteration of the parasitized crab. Insects are rarely hermaphroditic, but many species reproduce parthenogenetically (without fertilization). The insect ovary is composed of clusters of tubules (ovarioles) with no lumen, or cavity.

The upper portion of each ovariole gives rise to oocytes (immature eggs) that mature and are nourished by yolk from the lower portion. The oviduct leads to a genital chamber (copulatory bursa, or vagina), with which are often associated accessory glands and a seminal receptacle. Some accessory glands form secretions by which eggs become attached to a hard surface; others secrete a protective envelope around the egg. The eighth and ninth body segments are often modified for egg-laying. The paired testes consist of a series of seminal tubules that form primary spermatogonia (immature spermatozoa) at their upper ends. As the spermatogonia mature a covering is secreted around them. Eventually they enter a storage area (seminal vesicle). The terminal portion of the male system is an ejaculatory duct that passes through a copulatory organ. A pair of accessory glands, often associated with the ejaculatory duct, contributes to the semen (fluid containing sperm) or participates in spermatophore formation.

The ninth body segment and sometimes the tenth bear appendages for sperm transfer. Scorpions and spiders have tubular or saclike gonads; the female system is equipped to receive and store sperm, and, in some species, the female retains the eggs long after fertilization has occurred. Male spiders may have a cluster of accessory glands associated with the terminal portion of the reproductive system for the manufacture of spermatophores, or they may have expanded seminal vesicles for the retention of sperm until copulation takes place. Often specific appendages are adapted for sperm transfer.

REPRODUCTIVE SYSTEMS OF VERTEBRATES

GONADS, ASSOCIATED STRUCTURES, AND PRODUCTS

The reproductive organs of vertebrates consist of gonads and associated ducts and glands. In addition, some vertebrates, including some of the more primitive fishes, have organs for sperm transfer or ovipository (egg-laying) organs. Gonads produce the gametes and hormones essential for reproduction. Associated ducts and glands store and transport the gametes and secrete necessary substances. In addition to these structures, most male and female vertebrates have a cloaca, a cavity that serves as a common terminal chamber for the digestive, urinary, and reproductive tracts and empties to the outside. In lampreys and most ray-finned fishes in which the cloaca is small or absent, the alimentary canal has a separate external opening, the anus.

In some teleosts the alimentary, genital, and urinary tracts open independently. Hagfishes, which are closely related to the lampreys, have a short cloaca. In many vertebrates other than mammals, especially reptiles and birds, the cephalic, or head, end of the cloaca is partitioned by folds into a urinogenital chamber (urodeum) and an alimentary chamber (coprodeum) that open into a common terminal chamber (proctodeum). Above monotremes (*e.g.*, platypus, echidna) the embryonic cloaca becomes completely partitioned into a urinogenital sinus conveying urine and the products of the gonads, and an alimentary pathway; the two open independently to the exterior.

Gonads arise as a pair of longitudinal thickenings of the coelomic epithelium and underlying mesenchyme (unspecialized tissue) on either side of the attachment of a supporting membrane, the dorsal mesentery, to the body wall. At first, gonadal ridges bulge into the coelom and are continuous with the embryonic kidney. The germinal epithelium covering the gonadal ridges gives rise to primary sex cords (medullary cords) that invade the underlying mesenchyme. These cords establish within the gonadal blastema (a tissue mass that gives rise to an organ) a potentially male component, the medulla. Secondary sex cords grow inward, spreading just beneath the germinal epithelium to form a cortex. If the gonad is to become a testis, only the medullary component differentiates. If the gonad is to become an ovary, only the cortex differentiates.

The length of an adult gonad depends, in part, upon the extent of gonadal-ridge differentiation. In cyclostomes (lampreys and hagfish), elasmobranchs (sharks, skates, and rays), and teleosts most of it differentiates, and the gonads extend nearly the length of the body trunk. In tetrapods (amphibians, reptiles, birds, and mammals), the cranial portion, at the anterior end, generally does not differentiate; in toads only the more caudal, or posterior, portion does so. The middle segment in toads of both sexes gives rise to a Bidder's organ containing immature eggs. In anurans (frogs and toads) and some lizards of both sexes, one segment of the gonadal ridge gives rise to yellow fat bodies that, especially in anurans, diminish in size just prior to the breeding season. In mammals, only the middle portion of the gonadal ridge differentiates.

Some vertebrate species have only one gonad, which may lie in the midline or on one side; the condition is more common among females. Adult cyclostomes of both sexes have one gonad. In lampreys it is in the middle of the body; in hagfishes it is on the right side. Birds are the only other major group of vertebrates in which most females have one gonad, the right ovary being typically absent. Male birds have a pair of testes, however. Exceptions to the condition of single ovaries among birds include members of the falcon family, in which more than 50 percent of mature hawks have two well-developed ovaries. In all bird species a small percentage of females probably have two ovaries; reported instances include owls, parrots, sparrows, and doves, with estimates for doves ranging from 5 percent to 25 percent. A few teleosts and viviparous elasmobranchs have only one ovary; in sharks the right one is usually present, in rays, the left. In amniotes (*i.e.*, reptiles, birds, and mammals) unpaired gonads are unusual. Some lizards have one testis, and some female crocodiles have one ovary. Among mammals, the platypus usually has only a left ovary, and some bat species (family Vespertilionidae) have only the right.

One of two explanations may account for unpaired gonads: the paired embryonic gonadal ridges may fuse to form a median gonad—as in lampreys and the perch—or only one gonadal ridge may receive immigrating primordial germ cells (immature sperm or eggs), with the result that the opposite gonad does not develop—as in chickens and ducks. Both gonadal ridges have been reported to exhibit an equal number of primordial germ cells in embryonic hawks, and these typically have two ovaries.

Among lower vertebrates, mature gonads sometimes produce both sperm and eggs. Hermaphroditism is more general in cyclostomes and teleosts than in other fishes. A teleost may function as a male during the early part of its sexual life and as a female later. In some teleost families sperm and eggs mature simultaneously but in different regions of the gonad. These fish normally function as males during one season and as females the next. Cyclostomes generally are ambisexual during juvenile life—*i.e.*, immature male and female sex cells exist side by side, or, as in Myxine, the anterior part of the immature gonad may be ovary and the caudal part, the testis. It is thought that cyclostomes normally become unisexual at maturity. Hermaphroditism is uncommon among amphibians, although it frequently occurs as an anomaly. In vertebrates above amphibians, true hermaphroditism probably does not exist. Both male and female duct systems are occasionally absent. In cyclostomes, a few elasmobranchs, and some teleosts, such as salmon, trout, and eels, the gametes are propelled towards the posterior within the coelom, often by cilia (minute hairlike structures), and exit via a pair of funnel-like genital pores near the base of the tail. In cyclostomes, the pores lead to a sinus, or cavity, within a median papilla (*i.e.*, a fingerlike structure) and are open only during breeding seasons.

MALE SYSTEMS

Tests

In anurans, amniotes (reptiles, birds, and mammals), and even some teleosts, testes are composed largely of seminiferous tubules—coiled tubes, the walls of which contain cells that produce sperm—and are surrounded by a capsule, the tunica albuginea. Seminiferous tubules may constitute up to 90 percent of the testis. The tubule walls consist of a multilayered germinal epithelium containing spermatogenic cells and Sertoli cells, nutritive cells that have the heads of maturing sperm embedded in them. Seminiferous tubules may begin blindly at the tunic, or outermost tissue layer, and pass towards the centre, becoming tortuous before emptying into a system of collecting tubules, the rete testis.

Such an arrangement is characteristic of frogs. In certain amniotes—the rat, for example—the tubules may be open ended, running a zigzag course from the rete to the periphery and back again. The average length of such tubules is 30 centimetres (12 inches), and they seldom communicate with each other. In many mammals the tubules are grouped into lobules separated by connective-tissue septa, or walls. The arrangement permits the packing of an extensive amount of germinal epithelium into a small space. In immature males and in adult males between breeding seasons, the tubules are inconspicuous and the epithelium is inactive; in some species, however, spermatogenesis, or production of sperm, proceeds at a variable pace throughout the year. An active epithelium may exhibit all stages of developing sperm. The lumen, or tubule cavity, contains the tails of many sperm (the heads of which are embedded in Sertoli cells), free sperm, and fluid that is probably resorbed. In mammals, in any single zone along a tubule, all sperm are at the same stage of maturation; adjacent zones contain different generations of sperm, and a period of sperm formation and discharge is followed by an interval of inactivity.

In cyclostomes, most fishes, and tailed amphibians the germinal epithelium is arranged differently. Instead of seminiferous tubules there are large numbers of spermatogonial cysts (also called spermatocysts, sperm follicles, ampullae, crypts, sacs, acini, and capsules) in which sperm develop, but in which the epithelium is not germinal. Spermatogenic cells migrate into the cysts from a permanent germinal layer, which, depending on the species, may lie among cysts at the periphery of the testes or in a ridge along one margin of the testis. After invading the thin non-germinal epithelium of a cyst, spermatogenic cells multiply, producing enormous numbers of sperm. The cysts become greatly swollen and whitish in colour; the entire testis also swells and has a granular appearance. As sperm mature, they separate from the epithelium and move freely in the cystic fluid. Finally, the cysts burst, and the sperm are shed into ducts. In the case of cyclostomes and a few teleosts the sperm are shed into the coelom. The cysts, totally emptied, collapse. Then either they are replaced by new ones, or they become repopulated by additional spermatogenic cells. It is not yet known which of these processes occurs.

Testicular stroma, which fills the spaces between seminiferous tubules or spermatogenic cysts, consists chiefly of connective tissue, blood and lymphatic vessels, and nerves; it is more abundant in some vertebrates than in others. Glandular Leydig (interstitial) cells are also present in most, if not all, vertebrates. Thought to be a primary source of androgens, or male hormones, Leydig cells are not always readily distinguishable, and, in some bird species, they may be seen only with the electron microscope. The capillary system of the rat testis, and probably that of many other vertebrates, is such that blood that has bathed the Leydig cells flows to the tubules; it is thus probable that Leydig cell hormones have an immediate effect on the germinal epithelium.

Testes in vertebrates below mammals lie within the body. This is also true of many, sometimes all, members of the mammalian orders Monotremata,

Insectivora, Hyracoidea, Edentata, Sirenia, Cetacea, and Proboscidea. Some male mammals—most marsupials, ungulates, carnivores, and primates after infancy—have a special pouch (scrotum) that the testes occupy permanently. A few mammals have a pouch into which the testes descend and from which they can be retracted by muscular action. These include a few rodents such as ground squirrels; most, if not all, bats; and some primitive primates (loris, potto). The scrotum consists of two scrotal sacs, each connected to the abdominal cavity by an inguinal canal lined with the peritoneal membrane. The canals are the path of descent (and retraction) of the testes to the sacs.

In descending, the testes carry along a spermatic duct, blood and lymphatic vessels, and a nerve supply wrapped in peritoneum and constituting, collectively, the spermatic cord. Rabbits, most rodents, and some insectivores, which lack scrotal sacs, have instead a wide inguinal canal into which the testes may be drawn and from which they are retracted when in danger of injury. In these mammals, descended testes cause a temporary bulge in the perineal region (*i.e.*, between the anus and the urinogenital opening). In a small number of mammals, the testes permanently occupy the perineal location. The scrotum is a temperature-regulating device. Warm blood approaching the testis comes close to the vessels carrying cool blood leaving the testis, so that the blood approaching the testis is cooled; the vessels form an intricate vascular network (pampiniform plexus) within the spermatic cord. Failure of both testes to enter the scrotal sacs (cryptorchidism) results in permanent sterility. In cold weather two sets of muscles, the dartos and cremasteric, pull the testes close to the body.

The dartos lies between the two scrotal sacs and is attached to the scrotal skin. The cremaster, wrapped around the spermatic cord, is an extension of the abdominal wall musculature. It retracts the testis. Birds, like mammals, are homoiothermic (warm-blooded), and their testes are near air sacs (extensions of incurrent respiratory tubes). Air in the sacs may help regulate the temperature of the testes.

Ducts

The male duct system begins as the rete testis, a network within the testis of thin-walled ductules, or minute ducts, that collects sperm from the seminiferous tubules. The rete is drained by a number of small ducts—usually fewer than ten—called the vasa efferentia, which are modified kidney tubules. In some fishes and amphibians the vasa efferentia connect the testes with the cranial (anterior) end of the kidneys. In anamniotes (*e.g.*, fish and amphibians), therefore, except teleosts, the ducts that drain the kidneys usually drain the testes also. In most amphibians these ducts pass caudad, or posteriorly, to empty independently into the cloaca; in some fishes they pass through a median urinogenital papilla.

Although drainage of the testis and the kidney by the same duct is a basic pattern, there has been a tendency in many vertebrates towards separate spermatic and urinary ducts. This tendency is manifested in one of two ways among anamniotes. In many sharks and in some amphibians (Plethodontidae, Salamandridae, Ambystomatidae), the embryonic kidney duct ultimately drains the testis, and one or more new ducts (ureters) drain the adult kidney. On the other hand, in the primitive fish Polypterus and in most teleosts, the embryonic kidney duct drains the adult kidney, and a new duct arises to drain the testis. Many degrees of separation of the two ducts occur in anamniotes, from the condition of the sturgeon, in which the spermatic duct unites with the urinary duct far towards the head, to the condition in Esox (a pike), in which spermatic and urinary ducts empty independently to the exterior.

In amniotes, the mesonephric kidney is a temporary structure confined to the embryo, but the mesonephric duct persists in the adult male as a sperm duct. A separate ureter drains the adult kidney. The spermatic and urinary ducts empty independently into the cloaca except in mammals above monotremes, in which they are confluent with the urethra. The epididymis of amniotes, a highly tortuous duct draining the vasa efferentia, usually serves as a temporary storage place for sperm; it is small in birds and large in turtles. In mammals, the first part of the epididymis consists of a head, body, and tail that wrap around the testis; it gradually straightens to become the spermatic duct. The epididymis secretes substances that prolong the life of stored sperm and increase their capacity for motility.

In all vertebrates certain regions of the spermatic duct are lined by cilia and a variety of secretory epithelial cells. One end may enlarge to form a sperm reservoir or secrete seminal fluid. In the catfish Trachycorystes mirabilis secretions of the spermatic duct form a gelatinous plug in the female similar to the vaginal plug of mammals. A seminal glomulus in birds functions as a sperm reservoir. In some mammals an enlargement of the spermatic duct called the ampulla contributes to the seminal fluid and stores sperm. Small mucous glands (of Littré) and other glandular structures open into the urethra along its length. Cloacal glands in basking sharks and many salamanders form a jelly that encloses sperm in a spermatophore. Cloacal glands of some lizards produce secretions called pheromones. The siphon sac of elasmobranchs is one of the few accessory sex glands that is a separate organ in animals below mammals. It extends as an elongated pocket into the pelvic fin and secretes a nutritive mucus that enters the female reproductive tract with sperm.

ACCESSORY GLANDS

Accessory sex glands that are conspicuous outgrowths of the genital tract are almost uniquely mammalian. The major mammalian sex glands include the prostate, the bulbourethral, and the ampullary glands, and the seminal vesicles.

All are outgrowths of the spermatic duct or of the urethra and all four occur in elephants and horses and in most moles, bats, rodents, rabbits, cattle, and primates. A few members of these groups lack ampullary glands, or ampullary glands and seminal vesicles. Cetaceans (whales, porpoises) have only the prostate, as do some carnivores, including dogs, weasels, ferrets, and bears.

The prostate, the most widely distributed mammalian accessory sex gland, is absent only in Echidna (a marsupial) and a few carnivores. It empties into the urethra by multiple ducts. Many rodents, insectivores, and lagomorphs have three separate prostatic lobes; in a few mammals (some primates and carnivores) the prostate is a single mass with lobules and encircles the urethra at the base of the bladder. In a few mammals (*e.g.*, opossum), the prostate is not a compact mass but a partly diffuse gland. In many rodents (*e.g.*, rat, guinea pig, mouse, hamster) and some other mammals, the semen coagulates quickly after ejaculation as a result of a secretion from a male coagulating gland, which is usually considered part of the prostatic mass. Coagulated semen forms a vaginal plug that temporarily prevents copulation.

Bulbourethral (Cowper's) glands arise from the urethra near the penis and are surrounded by the muscle of the urethra or penis. Typically, there is one pair, but as many as three (marsupials) may be found. The glands, small in man, large in rodents, elephants, and some ungulates including pigs, camels, and horses, are absent in cetaceans, mustelids (*e.g.*, mink, weasel), sirenians (manatees, dugongs), pholidotans (pangolins), some edentates, and carnivores such as walrus, sea lion, bear, and dog.

Although many mammals have an ampullary swelling on the spermatic duct near the urethra, only a small number form a separate ampullary gland as an outgrowth of the duct. It is very large in some bats, absent in many mammalian orders, and variable in the rest. Although common in rodents, it is absent in guinea pigs and some strains of mice.

Seminal vesicles are paired, typically elongated and coiled fibromuscular sacs that empty into either the spermatic duct or the urethra. Absent in monotremes, marsupials, carnivores, cetaceans, and in some insectivores, chiropterans, and primates, seminal vesicles are exceptionally large in rhesus monkeys and small in man. They are absent in domesticated rabbits, small or rudimentary in cottontails, large in armadillos, and variable in sloths. They contribute the sugar fructose and citric acid to the semen but do not serve as sperm reservoirs.

FEMALE SYSTEMS

Ovaries

Ovaries lie within the body cavity and are suspended by a dorsal mesentery (mesovarium), through which pass blood and lymph vessels and nerves.

Primitive vertebrate ovaries occur in the hagfish, in which a mesentery-like fold of gonadal tissue stretches nearly the length of the body cavity. Unique in the hagfish is the fact that functional ovarian tissue occupies only the forward half of the gonadal mass, the rear part containing rudimentary testicular tissue. In most fishes except very primitive forms, the ovaries are similarly elongated. In tetrapods other than mammals, the ovaries are usually confined to the middle third or half of the body cavity, particularly during non-breeding seasons. The ovaries of mammals undergo moderate caudal displacement, finally coming to lie between the kidney and the pelvis.

The appearance of an ovary depends on many factors—*e.g.*, whether one egg or thousands are discharged (ovulated); whether the eggs are immature or ripe; whether mature eggs are small or large; or whether pigments occur in the egg cytoplasm, such as those responsible for yellow yolk. Other factors also affect the appearance of the ovary: the season of the year in seasonal breeders (the ovary enlarges during breeding seasons, diminishes in size between seasons); the age of the animal (whether juvenile, reproductively active, or senile, particularly in birds and mammals); and the fate of ovulated, or discharged, egg follicles, or sacs. The ovaries are covered with a germinal epithelium that is continuous with the peritoneum lining the body cavity. The term germinal epithelium is inappropriate because in most adults it contains no germ cells, these having moved deeper into the ovary. In hagfishes and amphibians, cells that give rise to eggs are known to occur in the germinal epithelium, and it may be that the germinal epithelium in a few other vertebrates contains similar cells. The germinal epithelium undergoes cell division, however. This is particularly true of species in which enormous expansion of the ovary occurs each breeding season. Beneath the epithelium is a layer of connective tissue, the tunica albuginea, which is much thinner than that surrounding the testes.

A typical vertebrate ovary consists of cortex and medulla. The cortex, immediately internal to the tunica albuginea, contains future eggs and, at one time or another, eggs in ovarian follicles (*i.e.*, developing eggs); it undergoes fluctuations in size and appearance that correlate with stages of the reproductive cycle. The cortex also contains remnants of ovulated follicles and, in mammals, clusters of interstitial cells that, in some species, are glandular. The cortical components are embedded in a supportive framework of connective, vascular, and neural tissue constituting the stroma. Internal to the cortex is the medulla, consisting of blood and lymph vessels, nerves, and connective tissue. The medulla, which contains no germinal elements, exhibits no significant cyclical activity, is usually inconspicuous, is continuous with the dorsal mesentery, and, in cyclostomes, is hardly distinguishable from the latter. The mammalian medulla, on the contrary, is almost completely surrounded by cortex and converges on the mesovarium (*i.e.*, the part of the peritoneum that supports the ovary) at a narrow hilus, at which nerves and vessels enter the ovary. In the medulla of the mammalian ovary near the hilus are small masses of blind tubules or solid cords—the rete ovarii—which are homologous (*i.e.*, of the same embryonic origin) with the rete testis in the male. The microscopic right ovary of birds usually consists only of medullary tissue.

Ovaries are characterized as saccular, hollow, lacunate (*i.e.*, compartmented), or compact. The ovary of many teleosts, especially viviparous ones, contains a permanent cavity, which is formed during ovarian development when an invagination of the ovarian surface traps a portion of the coelom. The cavity is therefore unique in that it is lined by germinal epithelium. The lining develops numerous ovigerous folds that project into the lumen and greatly increase the surface area for proliferation of eggs. In most other teleosts, a temporary ovarian cavity develops after each ovulation, when the shrinking cortex withdraws from the outside ovarian wall along one side of the ovary. The resulting cavity is obliterated as eggs of the next generation enlarge. The permanent and temporary cavities of teleost ovaries and a similar cavity in garfish ovaries are continuous with the lumen of the oviduct, and eggs are shed into them. The ovaries of other fishes lack cavities and are characterized as compact. The amphibian ovary, which contains six or more central, hollow sacs that give it a lobed appearance, is characterized as saccular. The sacs are formed when the embryonic medullary and rete cords become hollow and coalesce. Maturing eggs bulge into the sacs but are not shed into them. The ovaries of reptiles, birds, and monotremes have cavities homologous to those in amphibians; the number of medullary spaces in the adults is considerably larger, however, so that the ovaries contain an extensive network of fluid-filled cavities (lacunae). Such ovaries are characterized as lacunate. The ovaries of mammals above monotremes are compact, having no medullary cavities.

An ovarian follicle consists of an oocyte, or immature egg, surrounded by an epithelium, the cells of which are referred to variously as follicular, nurse, or granulosa cells. In cyclostomes, teleosts, and amphibians, the epithelium is one layer thick. In the hagfish and those vertebrates in which the oocyte receives heavy deposits of yolk (elasmobranchs, reptiles, birds, and monotremes), the epithelium appears to be two cells thick, apparently the result of layering of nuclei in a simple columnar epithelium (*i.e.*, epithelium consisting of relatively "tall" cells). Above monotremes the follicular epithelium appears to be many cells thick; in at least one species, however, this is considered an artifact, and all granulosa cells are said to extend between the outer boundary of the epithelium and the oocyte.

The follicular epithelium originates as a few flattened cells derived from the germinal epithelium. Primary follicles are usually situated just under the tunica albuginea; secondary follicles lie deeper in the cortex. The primitive role of the follicular cells appears to be the secretion of the yolk-forming material onto or into the oocyte. Evidence from mammals indicates that the follicular cells may also have a role in converting substances produced elsewhere into female hormones, or estrogens. In some hibernating bats the granulosa cells are filled with glycogen, or animal starch, which may be a source of energy. Mammalian follicles above monotremes are unique in that they develop a fluid-filled cavity (antrum) within the granulosa layer. During antrum formation cell division of the granulosa cells increases, and fluid-filled spaces develop among the cells.

The spaces coalesce to form the antrum. Under the influence of pituitary gonadotropic hormones, many antral follicles thereafter continue to grow, forming large so-called Graafian follicles—less than 400 microns, or 0.4 millimetre (0.16 inch), in diameter in large mammals, 150–200 microns, or 0.15–0.2 millimetre (0.006–0.008 inch), in small ones. Graafian follicles contain mature eggs and appear as large blisters on the ovary. At this stage the ovum, suspended within the fluid of the antrum (liquor folliculi) by a slender stalk of granulosa cells, is surrounded by a cluster of these cells, the cumulus oophorus, or discus proligerus. The remaining follicular cells form a thin wall surrounding the antrum.

Antra are lacking in a few insectivores (Hemicentetes, Euriculus) because the granulosa cells swell and multiply to form corpora lutea, masses of yellow tissue. In the bat Myotis the antrum is likewise compressed and disappears just before discharge of the egg, or ovulation. In all vertebrates, oocytes that have begun to grow and mature may, at any time until just before ovulation, cease development and undergo atresia, or degeneration. This is a normal process that reduces the number of eggs ovulated. In small laboratory rodents, atresia takes place in 50 percent of the Graafian follicles in each ovary one or two days before ovulation, thus reducing the number of ovulatable eggs by 50 percent. A similar reduction takes place in hagfish prior to ovulation. Atretic follicles eventually become lost in the stroma of the cortex of the ovary. In mammals especially, follicles lacking oocytes and antra, called anovular follicles, as well as polyovular follicles (*i.e.*, containing more than one oocyte), occasionally occur.

The ovarian follicle of vertebrates, commencing with hagfish, is surrounded by a theca, or sheath, composed of two concentric layers of stromal cells. The outer layer (theca externa) is chiefly connective tissue but may contain smooth muscle fibres. The inner layer (theca interna) has more blood vessels and, in vertebrates that produce heavily yolked eggs, the largest vessels carry venous blood. In these species the cell membranes of the oocyte and granulosa cells have many microvilli (*i.e.*, fingerlike projections), which probably facilitate transport of substances important in yolk formation from the blood vessels to the egg. Mature follicles in the marsupial Dasyatus are said to lack theca, and in some bats only one thecal layer has been described.

During the growth phase, eggs in species with massive amounts of yolk may increase in size 106 (1,000,000) or more times as a result of vitellogenesis (deposit of yolk). In goldfish, on the other hand, when vitellogenesis commences, the egg has a diameter of 150 microns (0.15 millimetre [0.006 inch]); that of the mature egg is only 500 microns (0.5 millimetre [0.02 inch]). Mammalian eggs contain little yolk and vary little in size. Oogonia (*i.e.*, cells that form oocytes) of the golden hamster average 15 microns (0.015 millimetre [0.0006 inch]) in diameter, and eggs in Graafian follicles average 70 microns (0.07 millimetre [0.003 inch]). The mature eggs of horses and humans are approximately the same size—somewhat less than 150 microns. In seasonally breeding oviparous fishes and amphibians, all eggs are usually in the same stage of development, and the ovary grows to a mature state quite rapidly as a result of growth of the eggs, which frequently number more than 1,000,000. Such ovaries distend the body wall when mature; following spawning, the ovaries shrink rapidly to inconspicuous bodies consisting mainly of oogonia, immature oocytes, and a few stromal cells.

In reptiles and birds, ovarian weight also is high in proportion to body weight during egg-laying seasons. The weight of the ovary of the starling, for example, may increase from eight milligrams in early winter to 1,400 milligrams immediately before ovulation. The mature eggs of reptiles and birds are unique in that they are suspended from the ovary by a short stalk (pedicle). The stalk contains a cortex with additional oocytes in various stages of development and extensions of vessels and nerves. Full growth of the follicle in reptiles and birds requires only a few days or weeks (nine days in the domestic hen). In mammals, the ratio of ovarian weight to body weight varies insignificantly throughout the reproductive life of the female, and follicles in many stages of development are constantly present.

Vertebrate eggs are almost universally shed into the coelom or into a subdivision thereof, from which they enter the female reproductive tract. Even in those teleosts in which the eggs are shed into an ovarian cavity, the latter is often of coelomic origin. In many mammals a membranous sac of peritoneum, the ovarian bursa, traps part of the coelom in a chamber along with the ovary. The bursal cavity (periovarian space) may be broadly open to the main coelom, completely closed off from the coelom, or in communication with the coelom by a narrow, slitlike passage. The bursa, moderately developed in lower primates and catarrhines (Old World monkeys), is poorly developed in man. In horses, one edge of the ovary contains a long groove (ovulation fossa) into which all eggs are shed; the groove is found in a cleftlike ovarian bursa. The ovarian bursa increases the probability that all ovulated eggs will enter the oviduct.

The process of ovulation has been described for all vertebrate classes. Elasmobranchs, reptiles, and birds have massively yolked eggs. As ovulation approaches, the fimbria (*i.e.*, frills, or fringes) of the membranous and muscular funnel surrounding the entrance to the oviduct wave in a gentle, undulating motion. An egg that is nearly free of the ovary is grasped and partially encompassed by the fimbria; when the egg is freed, the fimbria draw the egg into the funnel. At this time, the egg has little shape and is partly squirted and partly flows into the oviduct; never completely free in the coelom, its chances of not entering the oviduct are small. In the case of moderately or poorly yolked eggs cilia help to sweep the eggs into the ostium, or opening, of the oviduct. During ovulation in Japanese rice fish, Oryzias latipes, a tiny papilla, or fingerlike process, develops on the surface of a bulging mature follicle in the centre (stigma) of the follicle. The follicle becomes thin at the stigma, an aperture appears, and the egg rolls out. In rabbits this process differs only in detail. During the final 20 minutes before ovulation in rabbits, some of the tiny blood vessels surrounding the stigma rupture, and a small pool of blood forms under the apex of the cone-shaped papilla. The follicular wall shortly gives way at the apex, and follicular fluid oozes from the opening, followed soon after by the egg. The ovulated mammalian egg typically is surrounded by a layer of columnar follicular cells, the corona radiata; but it is naked in some insectivores and some marsupials. Following ovulation in all vertebrates, the ovary may become smaller, become modified for maintenance of pregnancy, or proceed to form additional eggs.

The process of ovulation in vertebrates has been documented, but the immediate causes remain to be clarified. It is almost certain that an ovulatory hormone is secreted by the pituitary gland (*i.e.*, the so-called master endocrine gland) of all vertebrates. It is highly probable that breakdown of very small fibres that bind the follicular cells together may occur at the stigma, weakening the follicular wall at that location. Hormones from the ovary and other sources may play a role, as may neurohormones, which are hormones released at nerve endings. Rhythmic contractions of the entire ovary occur at ovulation in many vertebrates and have been described in rabbits. The role of mechanical pressure within the follicle, however, is not understood. Ovulation in most mammals (spontaneous ovulators) occurs cyclically as a result of the spontaneous release of the ovulatory hormone. In a few mammals (reflex ovulators) the stimulus of copulation is essential for release of the ovulatory hormone.

Striking postovulatory changes take place in the follicles of mammals and, to lesser degrees, of lower vertebrates. Blood vessels from the theca interna invade the ovulated follicles; the granulosa cells divide, enlarge, accumulate fats, and obliterate any remnants of the collapsed antra. Thereafter, they are known as lutein cells. Theca interna cells undergo changes identical to those of the granulosa cells.

The result in mammals is the formation of solid masses called corpora lutea, recognizable as prominent reddish-yellow bulges on the ovary. Corpora lutea produce the hormone progesterone, which is essential for the maintenance of pregnancy. The conversion of postovulatory follicles into structures more or less resembling mammalian corpora lutea has been demonstrated in numerous viviparous reptiles, amphibians, and elasmobranchs; in certain other fishes, including cyclostomes; and in some oviparous amphibians and reptiles. In birds, the postovulatory follicle shrinks, and identifiable corpora lutea do not develop, although some granulosa cells accumulate lipids of unknown significance.

TRACTS

The female reproductive tract consists of a pair of tubes (gonoducts) extending from anterior, funnel-like openings (ostia) to the cloaca, except as noted below. The gonoducts are specialized along their length for secretion of substances added to the eggs; for transport, storage, nutrition, and expulsion of eggs or the products of conception; and, in species with internal fertilization, for receipt, transport, storage, and nutrition of inseminated sperm. The predominately muscular tracts are lined by a secretory epithelium and ciliated over at least part of their length. Fusion of the caudal (tail) ends of the paired ducts may occur. Gonoducts are absent in cyclostomes and a few gnathostome fishes that have abdominal pores. A few vertebrates have only one functional gonoduct.

Gonoducts in lungfishes and amphibians are coiled muscular tubes that are ciliated over most of their length. Only occasionally do they unite caudally in a genital papilla before opening into the cloaca. During breeding seasons their diameter increases severalfold because of the highly active secretory epithelium. Between breeding seasons they are small. In some anurans (frogs, toads), such as Rana, the lower end of each gonoduct is expanded to form an ovisac, in which ovulated eggs are stored until spawning; the tube between the ostium (funnel-like opening) and ovisac is the oviduct. In viviparous amphibians the young develop in the ovisac. In amphibians, numerous multicellular glands extend deep into the lining of the female tract. Six successive glandular zones have been described in some urodeles, and these secrete six different gelatinous substances upon the egg. Female urodeles often have convoluted tubular outpocketings of the cloaca called spermatheca; they temporarily store sperm liberated from the male spermatophore.

The two gonoducts of elasmobranchs share a single ostium, a trait found only in Chondrichthyes. The ostium is a wide caudally directed funnel supported in the falciform ligament, which is attached to the liver. The role of the fimbria of the ostium at ovulation has been described. Two oviducts pass forward from the ostium to the septum transversum (*i.e.*, between the heart and abdominal cavities), curve around one end of the liver, then pass posteriorly on each side. Approximately midway between ostium and uterus each oviduct has a shell (nidamental) gland.

Fertilization takes place above the shell gland, which may be immense or almost undifferentiated. Half of the shell gland secretes a substance high in protein content (albumen), and the other half secretes the shell—delicate in viviparous forms, thick and horny in most oviparous species. Horny shells may have spiral ridges and many long tendrils, which entwine about an appropriate surface after the egg is deposited. In the viviparous shark Squalus acanthias several eggs pass one after the other through the shell gland, where they are enclosed in one long delicate membranous shell that soon disintegrates. Beyond the shell gland the oviducts terminate in an enlargement, which, in viviparous species, serves as a uterus. An oviducal valve may be found at the junction of oviduct and uterus. Although the two uteri usually open independently into the cloaca, they occasionally unite to form a bicornuate (two-horned) structure. In immature females, the uterus may be separated from the cloaca by a hymen, or membrane. The tract enlarges enormously during the first pregnancy and does not thereafter fully regress to its original size.

The gonoducts of most lower ray-finned fishes resemble those of lungfish, but those of gars and teleosts are exceptional in that the oviducts are usually continuous with the ovarian cavities. A median genital papilla receives the oviducts in teleosts, and the papilla is sometimes elongated to form an ovipositor. European bitterlings deposit their eggs in a mussel by means of the ovipositor, and female pipefish and sea horses deposit them in the brood pouch of a male. With certain modifications, the gonoducts of reptiles and birds are comparable to those of lower vertebrates. Crocodilians, some lizards, and nearly all birds have one gonoduct; the other is not well developed. Even in birds of prey having two functional ovaries, the right oviduct is sometimes undeveloped. The tracts of reptiles generally show less regional differentiation than do those of birds.

The oviduct funnel (ostium) in birds forms the chalazae—two coiled, springlike cords extending from the yolk to the ends of the egg. In both reptiles and birds, much of the length of the female tract is oviduct. This region, called the magnum in birds, secretes albumen; lizards and snakes do not form albumen. Behind the albumen-secreting region is a shell gland. In lizards, the gland is midway along the tract. In birds, the shell gland is at the posterior end, has thick muscular walls, and is often inappropriately called a uterus. It is preceded by a narrow region, or isthmus, which secretes the non-calcareous, or soft, membranes of the shell. The shell gland leads to a narrow muscular vagina that empties into the cloaca. The vagina secretes mucus that seals the pores of the shell before the egg is expelled. Special vaginal tubules (spermatheca) store sperm over winter in some snakes and lizards; seminal receptacles have been described in the oviduct funnel in some snakes. In birds, sperm storage glands (sperm nests) often occur in the funnel and at the uterovaginal junction. In lizards and birds, ovulation does not usually occur into a tract already containing an egg. Some lizards shed very few eggs per season; the gecko, for example, sheds only two.

The female reproductive tracts of monotremes, the egg-laying mammals, consist of two oviducts, the lower ends of which are shell glands. These open into a urinogenital sinus, which, in turn, empties into a cloaca. Marsupials have two oviducts, two uteri (duplex uterus), and two vaginas. The upper parts of the vaginas unite to form a median vagina that may or may not be paired internally. Beyond the median vagina, the vaginas are again paired (lateral vaginas) and lead to a urinogenital sinus. The posterior end of the pouchlike median vagina is separated from the forward end of the urinogenital sinus by a partition. When the female is delivering young, the fetuses are usually forced through the partition and into the urinogenital sinus, bypassing the lateral vaginas.

The ruptured partition may remain open thereafter, resulting in a pseudovagina. It closes in opossums, and in kangaroos both the median and lateral routes may serve as birth canals. The lateral vaginas in marsupials receive the forked tips of the male penis. Fertilization in all mammals takes place in the oviducts (Fallopian tubes). In eutherian mammals (*i.e.*, all mammals except monotremes and marsupials), with exceptions noted below, female reproductive tracts beyond the ostia (oviduct funnels) consist of two narrow and somewhat tortuous Fallopian tubes, two large uterine horns (each of which receives a Fallopian tube), a uterine body, and one vagina. Fallopian tubes often have a short dilated ampulla, or saclike swelling, just beyond the ostium. Implantation of the egg occurs only in the uterine horns; the embryos become spaced equidistant from one another in both horns even if only one ovary has ovulated. In some species one horn is rudimentary—the left in the impala (an African antelope)—and the embryos become implanted in the other horn, even though both ovaries ovulate. The body of the uterus in some mammals (*e.g.*, rabbits, elephants, aardvarks; some rodents, bats, insectivores) contains two separate canals (bipartite uterus).

In other mammals (ungulates, many cetaceans, most carnivores and bats) the body of the uterus has one chamber into which the two horns empty (bicornuate uterus). There are numerous intermediate conditions between the bipartite and bicornuate condition. Apes, monkeys, and man have no horns, and the Fallopian tubes empty directly into the body of the uterus (simplex uterus). In all mammals, the uterine body tapers to a narrow neck (cervix). The opening (os uteri) into the vagina is guarded by fleshy folds (lips of the cervix). The vagina in eutherian mammals other than rodents and primates terminates in a urinogenital sinus that opens to the exterior by a urinogenital aperture. In some rodents and in higher primates the vagina opens directly to the exterior. In the young of many species a membrane, the hymen, closes the vaginal opening. In guinea pigs the hymen reseals the opening after each reproductive period. Sperm are stored over winter in the uterus of some bats and in vaginal pouches in others.

IN VITRO FERTILIZATION AND EMBRYO TRANSFER

Insemination is done by mixing about 1 ml culture medium containing oocytes with 1 ml of semen preparation (that contains 10,000-50,000 spermatozoa/ml). Oocytes are incubated for 12-13 h. Thereafter, they are observed keeping them in a Petri dish for the presence of two pronuclei and two polar bodies. Besides, the abnormal fertilized oocyte contains more than two pronuclei and granulation in cytoplasm. First cleavage occurs after 24-30 h of insemination. Delay in cleavage after insemination confirms that the embryo is abnormal. Thus, the abnormal embryo is discarded; only normal embryo is used. After fertilization, embryo should not be detained for el long time in vitro. It should be transferred to treated recipient female when it is in 2-4 cell stage. However, success has been achieved with transfer of 1-16 celled embryos. However, early embryo fails to survive in uterus. This stage is very critical, therefore, much care should be taken.

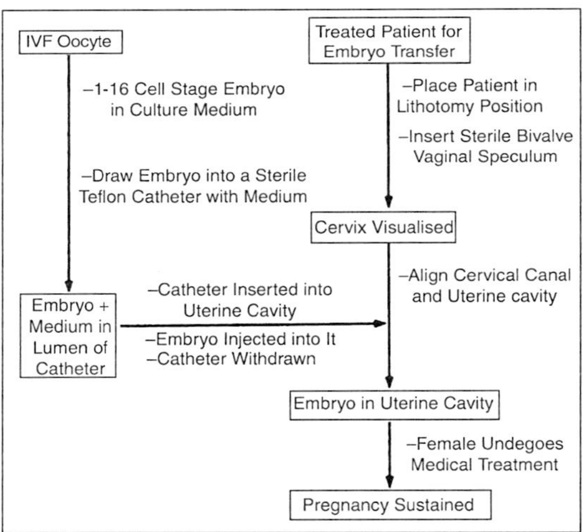

Fig. Outline of Embryo Transfer in Treated Female Human.

Some times, multiple embryo transfer is also done. It may increase the chance of embryo survival, but results in multiple pregnancy to which the female does not like. The transferred embryo stays in fallopian tube up to 8-16 cell stage, thereafter moves to uterus. Embryo of prolonged stage in vitro should not be transferred because of less chance of survival and pregnancy.

Embryo transfer should be done in an operation theatre. No anaesthesia should be given to female. She is given only Diazepam (Valium) of 10 mg orally before embryo transfer. The embryo is to be transferred through cervical canal; therefore, the patient is placed in lithotomy (*i.e.* in the position of knee and chest with tilting the head down) so that the fundus should be in lower position than cervical canal. The cervix is visualized by inserting a sterile bivalve vaginal speculum. Then uterine, cavity is aligned to cervical canal by manipulating the uterus. Distance between fundus and external cervical is earlier measured through ultrasonography. The embryo is drawn into a teflon catheter along with 10 ml of tissue culture medium. The catheter is gently inserted into culture medium. It should be made sure that the embryo has been transferred or not. The catheter is observed under microscope. Absence of embryo in catheter ensures for its transfer in uterus. Wrong placement of catheter or high amount of medium will result in tubal pregnancy or expulsion of embryo from uterine cavity. Therefore, care should be taken for proper handling. After embryo transfer, the female is allowed to take rest for about 2-7 h and abstain from intercourse for about 10 days. The patient follows the advice of doctor and undergoes for regular check up.

IN VITRO FERTILIZATION (IVF) TECHNOLOGY

The term in vitro means in glass or in artificial conditions, and IVF refers to the fact that fertilization of egg by sperm had occurred not in uterus but out side the uterus at artificially maintained optimum condition. In recent years the IVF technology has revolutionized the field of animal biotechnology because of production of more and more animals as compared to animal production through normal course.

For example, an animal produces about 4-5 offsprings in her life through normal reproduction, whereas through IVF technology the same can produce 50-80 offsprings in her life. Therefore, the IVF technology holds a great promise because a large number of animals may be produced and gene pool of animal population can also be improved. In India M.L. Madan, an animal embryo-biotechnologist at National Dairy Research Institute, Karnal (Haryana) has got success in producing more calves in cows. The IVF technology is very useful.

It involves the procedure:

- Taking out the eggs from ovaries of female donor,
- In vitro maturation of egg cultures kept in an incubator,
- Fertilization of the eggs in test tubes by semen obtained from superior male,
- Implantation of seven days old embryos in reproductive tract of other recipient female which acts as foster mother or surrogate mothers.

These are used only to serve as animal incubator and to deliver offsprings after normal gestation period. The surrogate mothers do not contribute any thing in terms of genetic make up since the same comes from the egg of donor mother and semen from artificial insemination.

In Vitro Maturation (IVM) of Oocytes

The immature oocytes are incubated in vitro so that they can be mature. However, immature oocytes should be taken out from follicles because they cannot mature in it but degenerate. Therefore, full potential of superovulation and all the oocytes can be utilized by IVF technology. Moreover, metabolic and hormonal requirement for oocytes during IVM should be found out so that the present rate of maturation (20%) could be improved. In majority of cases ovarian follicles never reach maturity and degenerate due to unknown causes. Possibly there may be genetic defects associated with them.

Culture of in Vitro Fertilized Embryos

IVF ofeggs is carried out in small droplets (microdroplets) of culture medium. Each microdroplet comprises of about 10 oocytes. The medium should be supplemented with penicillamine, hypotaurin, and epinephrine because they facilitate penetration of sperms into oocytes. Moreover, one dose of sperm is given that consists of about one million sperms per ml of medium. Thus, IVF embryo must be maintained at in vitro conditions for a few days so that it may develop into blastocyte. It takes about seven days for sheep and goats and eight days for cattle. There are many laboratories where about 60 per cent IVF embryos of catties are cultured to blastocyte stage.

The term delivery from cultured embryo is very low due to occurrence of high loss during first two months of pregnancy. This may be due to abortion of foetuses arising from the presence of genetic defects. It should be noted that before birth about 80 per cent genes play a key role in differentiation and development of foetuses. The oocytes which are forced to mature in vitro occasionally bears some defects. Some times environmental mutagenesis occurs in eggs, sperms or embryos. Artificial culture media should be improved as oxygen may have toxic effect. Therefore, gas atmosphere should be carefully controlled.

The other most successful method of IVM is to place the fertilized zygotes into agar (so that it may wrap around it) and implant them in oviduct of synchronized sheep or rabbit where the environment for early development of embryo is perfect. For early bovine embryo the oviduct of rabbit and sheep has been used as in vitro culture system. Hundreds of cattle eggs can be put into oviduct of a sheep and many of these are recovered after a week. A good quality of embryo at the late morula/blastocyte stage of development with a yield of about 40 per cent or more has been recorded by Lu *et al*.. Brackett *et al*. reported the birth of first IVF calf after getting success in fertilizing the eggs recovered from ovulated cow. Thereafter, hundreds of IVF calves have been born in Japan, India, U.K., etc.

EMBRYO CLONING

A clone is a population of cells or organisms derived asexually from a single ancestor. They are genetically identical to each other to their common ancestor. Cloning means the production of exact genetic replica copies of an individual.

They can not be considered as an offspring but simply the copy of a given individual. Much work has been done on cloning in plants and microorganisms. However, the techniques used in plants can not be applied for animals. Moreover, many animals from a single genetically superior embryo can be produced. Still there is no method of finding out which embryos are capable of cloning. It is useless to clone an embryo if it is not superior.

Quadriparental Hybrids

During 1960s, Beatrice Mintz at Cancer Research Institute, Philadelphia (USA) demonstrated the interesting experimentation. She carried out fusion of embryos of two different species of mouse. This resulted in the formation of a single embryo which finally developed into a normal healthy animal having four parents.

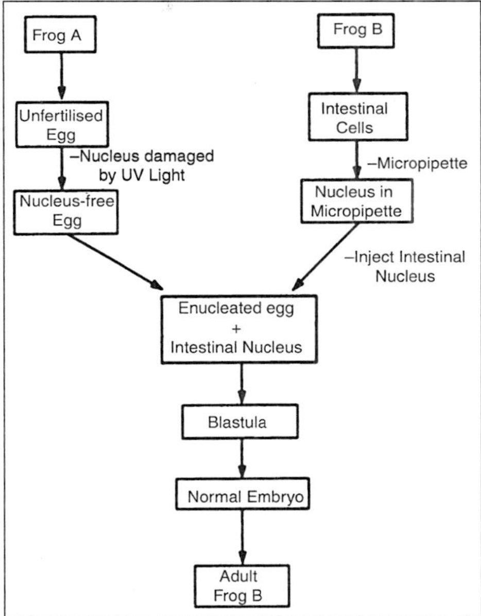

Fig. Nuclear Transplantation in Frog (Diagrammatic).

Embryo A was derived from the cross of male x female of one species, and embryo B derived after cross of male x female of the other species. The embryo A and B were united together and produced a single mass. In this experiment Mintz removed zona pellucida membrane of two early embryos and placed them in a suitable culture medium. The embryonic cells of the two embryos of blastula stage united randomly into a single mass of double sized embryo (blastocyst). A fresh membrane developed around the embryo. Then the embryo was transferred into the uterus of a foster mother. The foster mother was mated with a sterile male to bring her into proper stage for implantation. The first offspring with four parents was born in 1965. Similarly, exciting experiments have been done on another mammal. Following the same technique a hybrid of goat and sheep named geep was produced. At present two types of techniques for embryo cloning *viz.*, nuclear transplantation (transfer) and embryonic stem cells, are being developed.

Nuclear Transplantation

Nuclear transplantation (also nuclear transfer) involves removal of a single blastomere from a cleavage stage embryo with a fine micropipette of glass, and placing it under the outer membrane of an unfertilized mature enucleated oocyte (whose haploid nucleus has been removed by using micropipette or destroyed by UV light). For the first time in 1955, Robert Briggs and Tom King at Cancer Research Institute, Philadelphia (USA) carried out nuclear transplantation experiment on embryonic cells of frog. They transferred nucleus of undifferentiated blastula (a stage soon after fertilization of egg) into an enucleated egg cell. They noticed the normal development of the embryo.

When they performed serial transplantation of differentiated nucleus from late gastrula (a stage after blastula) into a nucleus-free unfertilized egg, abnormal embryos were formed. This shows that cell nucleus is differentiated with embryo development. In 1960s, J.B. Gurdon at Oxford University, U.K. transferred differentiated intestinal nucleus of a frog into nucleus-free unfertilized egg of different amphibian species (Xenopus laevis). The embryo developed into tadpole and matured into frog. This new enucleated cell developed into normal embryo. Any damage to the donor nucleus during transplantation leads to abnormal development.

DOLLY-The First Mammalian Clone

'Dolly', the worlds' first mammalian clone has been created from a fully differentiated non-reproductive cell of an adult sheep. It was born in February, 1996. The name Dolly has been given after an American country singer, Dolly Parton. In 1995, Ian Wilmut and his team of researchers at Roslin Institute, Edinburgh, Scotland, took udder (a fully differentiated tissue) from six year old sheep, Fin Dorset Ewe, and placed it in special solution that controlled cell cycle of cell division.

The cell was deprived off certain nutrient. At the same time an unfertilized egg was obtained from another adult sheep. Its nucleus was carefully removed leaving the intact cytoplasm in egg. The nucleus of udder cell was taken out and transferred into nucleus-free egg. This was facilitated by applying mild electric sock. The newly transplanted nucleus soon became functional according to the new cytoplasm in which it had been artificially transferred. This viable combination underwent cleavage like normal zygote. This so called embryo was then transplanted into the uterus of a third adult sheep (surrogate mother/ foster mother) for its further development.

Finally, a normal healthy little lamb, Dolly was born in February, 1996 which was genetically similar to the clone mother from which nuclear DNA was taken out. It does not have any similarity with that sheep from which egg was taken out or surrogate mother because they did not contribute any genetic character. Thus, Dolly has only a single parent because she has born asexually, a characteristic feature found in lower forms of animal life, not in mammals.

Although behind this great success the rate of success is very slow, yet it has given some hope to embryo-biotechnologists to bring about refinement. Out of 277 nuclei transferred singly to enucleated egg, only 29 eggs grew into embryos. Out of these, only 13 embryo could be successfully transplanted into surrogate mothers. Of these only one ewe was successful in giving birth to an offspring, Dolly.

The significant considerations that can be derived from this experiment are that:

- The genes of differentiated cells have inherent totipotency,
- The interplay between the regulatory system of the genome in nucleus and the cytoplasmic factors of egg may make a cell totipotent,
- Possibly the cytoplasm of enucleated egg makes the transplanted nucleus totipotent like that of normal fertilized egg nucleus,
- The maternally derived information in egg cytoplasm has an important role in cleavage which usually occurs after fertilization.

Hence, it is the egg cytoplasm but not the nucleus that regulates cleavage. Because the udder cell nucleus has limited potential for mitosis; it is the egg cytoplasm that interacted intracellularly with udder cell nucleus and stimulated to undergo repeated mitosis,

- Carbon copy of the adult sheep could be produced without involving sperms from male partner,
- The cloned animal produced via nuclear transplan-tation technique will be capable of restoring fertility as in 1998 Dolly gave birth to a little lamb named Bonny.

Embryonic Stem (ES) Cells

Cloning of mice could not be done as in sheep via nuclear transplantation. This was due to acceleration of developmental programmes of embryo. However, it is evident that before first embryonic division the cell has started its process of differentiation.

Therefore, for cloning of mice an alternative approach has been made *i.e.* the use of ES cells. A blastocyst of mouse is place in culture condition. The inner cells that form future foetus continue to divide and remain in undifferentiated totipotent state as ES cells. There is a peptide growth factor known as leukaemia inhibitory factor (LIF) which establishes and maintains ES cell lines.

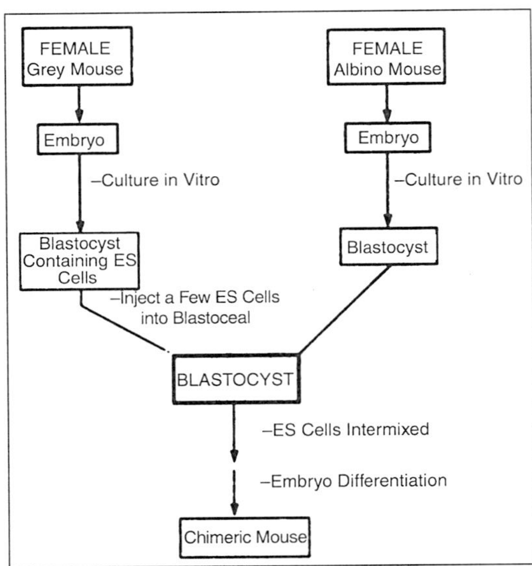

Fig. Production of Chimeric Mouse by Embryonic Stem Cell Transplantation.

The ES cell lines will be very useful in the area of production of transgenic animals. However, the ES cells are used in two different ways: a small number of ES cells can be injected into blastocoel space of a blastocyst. The ES cells get mixed with inner mass of cells of blastocyst to produce a chimera mouse which is a mixture of two cell genotype having the patches of different coloured fur. Crossing of male and female chimera will allow selection of homozygous mice derived from ES cells.

In Vitro Fertilization and Embryo Transfer in Humans

Due to several complicacies associated with humans, there has been the birth of in vitro fertilization technology. The IVF technology is a boon to childless couples. Initially this technique was pioneered by Prof. Robert Winston during 1970s which was later applied by P. Steptoe and R. Edwards for the production of first test tube baby, Louse Joy Brown, who was born on July 25, 1978. Since then more than 25,000 test tube babies have been produced so far through out the world.

5

Food of Transgenic Animals and Derived Products

Concern about the quality and safety of food products has been growing since the early 1980s. Many hazards have been identified, studied and handled by instituting national regulations and international standards and by setting up regulatory and inspection systems in processing plants, livestock farms and slaughterhouses.

These hazards can be biological (bacteria, parasites, viruses, prions), physical (metals, etc.) and chemical (residues of pesticides or veterinary drugs, toxic substances, etc.). New hazards are now perceived in relation to the use of transgenic animals. Transgenic animals include those obtained through nuclear transfer, those with improved production traits and those developed for biomedical purposes. Animals for xenotransplantation are barred from the food chain by national regulations.

PRODUCTS DERIVED FROM CLONED ANIMALS

In 2002, some 1400 Holstein cattle produced by blastomere nuclear transfer (not somatic cells) were raised in the US. They and their subproducts (animal and milk) are marketed and enter the food chain. No result of analysis of products derived from these transgenic animals is so far available, nor is there any information on abnormalities of gene expression in these adult cloned animals or analytical information on the composition of milk or meat derived from them or their offspring.

The animals are considered perfectly safe and no apparent ill effects have been noted among consumers. However, caution and risk analysis are recommended for such derived products, especially in the light of the problem of reprogramming the inserted nucleus. The safety of food from offspring of cloned animals is a controversial subject. Some see no risk as they originate from natural mating, while others point out that the offspring of a cloned animal has one half of the latter's genome. The cloning of animals from somatic cells is a recent technique, so it is difficult to make a firm decision on the safety of related milk or meat. Genetic variations appearing with nucleus differentiation and possibly causing modified genome expression are not known. The findings of studies on products of cloned animals will no doubt be available soon.

UNLOCKING THE SECRETS OF FUNCTIONAL FOOD COMPONENTS

Food technology and improved nutrition have played critical roles in the dramatic increase in life expectancy over the past 200 years, but the impact of diet on health is much broader than basic nutrition. A growing body of evidence documents positive health benefits from food components not considered nutrients in the traditional definition.

Scientific advances have allowed researchers to better characterize the biological basis of disease states, understand the metabolism of food at the cellular level, and identify the role of bioactive components in food and assess their impact on metabolic processes. New powerful analytical tools can enable scientists to unlock the biological functions of vast numbers of food components and their role in disease prevention and health promotion.

Functional foods can take many forms. Some may be conventional foods with bioactive components that can now be identified and linked to positive health outcomes. Some may be fortified or enhanced foods, specifically created to reduce disease risk for a certain group of people. Consumers can already select from a wide spectrum of foods that contain functional components either inherently or via fortification.

Health benefits may result from increasing the consumption of substances already part of an individual's diet or from adding new substances to an individual's diet. As additional bioactive components are identified, the opportunities for developing functional foods will be broad. Foods that naturally provide a bioactive substance may be enhanced to increase the level present in the food.

Alternately, foods that do not naturally contain a substance can be fortified to provide consumers with a broader selection of food sources for a particular component and its health benefit. Areas for research include better understanding the role and optimal levels of traditional nutrients for specific segments of the population, as well as identifying bioactive substances present in foods and establishing optimal levels. Early nutrition research focused on the range of vitamin and mineral intakes necessary to prevent frank deficiencies.

Now, researchers are investigating the optimum intake levels for traditional nutrients and the differences for various subpopulations. Understanding the role of nutrients at the molecular level will result in even more specific recommended dietary allowances for different population subgroups. Similar research is needed to identify the role of other bioactive food components, an area of research that is still in its infancy.

Only recently, several government agencies have begun developing a standard definition for "bioactive" food components. Research has proven that food and isolated food components can reduce the risk of disease, from the effect of vitamin A from eggs on blindness to the effect of zinc from high-protein foods on the immune system. Some examples of foods that may be considered functional foods include calcium-fortified orange juice, phytosterol/stanol-fortified spreads and juices, folate-enriched foods, soluble oat fibre, cranberry, and soy.

Table. Examples of Functional Food Components Currently Marketed

Functional Component Status of Claims	Health Benefits	U.S. Regulatory
Soluble oat fibre health claim	Coronary heart disease	FDA approved
Soy protein health claim	Coronary heart disease	FDA approved
Phytosterol/stanol esters health claim	Coronary heart disease	FDA approved (interim final rule)
Calcium health claim	Osteoporosis	FDA approved
Folate-enriched foods health claim	Neural tube defects	FDA approved

Research currently underway at academic, industry and government facilities will reveal how a myriad of substances can be used as functional food components. Although additional research is necessary to validate efficacy and establish appropriate dietary levels, researchers have identified functional food components that may improve memory, reduce arthritis, reduce cardiovascular disease and provide other benefits typically associated with drugs. In addition, new technologies will provide opportunities to produce bioactive food components from nontraditional sources. For example, Abbadi developed transgenic plant oils enriched with very long chain polyunsaturated fatty acids.

Other research has produced stearidonic acid in canola seeds to provide another source of omega-3 fatty acids in the diet. Emerging science requires that we broaden our frame of reference to take full advantage of these new discoveries. Foods may be developed to promote the expression of specific metabolites, reducing or preventing common diseases that afflict consumers with a specific genotype. Consumers might select functional foods and tailor their diets to meet changing health goals and different requirements at different ages. Future benefits might include functional foods for increased energy, mental alertness, and better sleep.

SHIFTING THE PARADIGM FOR HEALTH AND WELLNESS

A growing number of consumers perceive the ability to control their health by improving their present health and/or hedging against aging and future disease. These consumers create a demand for food products with enhanced characteristics and associated health benefits. In one study, 93% of consumers believed certain foods have health benefits that may reduce the risk of disease or other health concerns.

In addition, 85% expressed interest in learning more about the health benefits offered by functional foods. Using foods to provide benefits beyond preventing deficiency diseases is a logical extension of traditional nutritional interventions. Nonetheless, such an extension requires changes in not only the foods themselves, but also their regulation and marketing—truly a paradigm shift. Creating a scientifically valid distinction between food and medicine has never been easy. Centuries ago, Hippocrates advised, "Let food be thy medicine and medicine be thy food." Early nutrition research resulted in cures for numerous widespread deficiency-based diseases. Recent scientific advances have further blurred the line between food and medicine, as scientists identify bioactive food components that can reduce the risk of chronic disease, improve quality of life, and promote proper growth and development.

The Traditional Paradigm

Traditional fortification of foods with vitamins and minerals has been accepted by consumers and regulators, but consumers should recognize the clear distinction between the use and purpose of foods vs. drugs.

Table. Benefits and Risks of Foods vs. Drugs

Food and Food Components	Drugs
Energy/nutrition/necessary for life	Treatment of disease
Life long use and benefits	Immediate effect
All populations	Target population
Safea[a]	Benefit > risk
Consumer selects	Health provider prescribes

Notes:

Adapted from Yetley, 1996

[a]Safe when consumed as a food, but with a potential increase in risk as the component levels increase. Safety evaluation will be conducted to identify the limits

Food has traditionally been viewed as a means of providing normal growth and development. Regulatory policies were established to replace nutrients lost during processing and, in some cases, to prevent nutrient deficiencies in the population. Federal policies have generally required that other diseases be treated and managed through the use of drugs.

A New Paradigm

A new self-care paradigm recognizes that foods can provide health benefits that can co-exist with traditional medical approaches to disease treatment.

Science has clearly demonstrated additional dietary roles in reducing disease risk, and consumers have learned that food has a greater impact on health than previously known. At the same time, consumers recognize problems with the current healthcare system, perceiving that it is often expensive, time-constrained, and impersonal. Functional foods fit into a continuum that ranges from health maintenance/promotion to disease treatment.

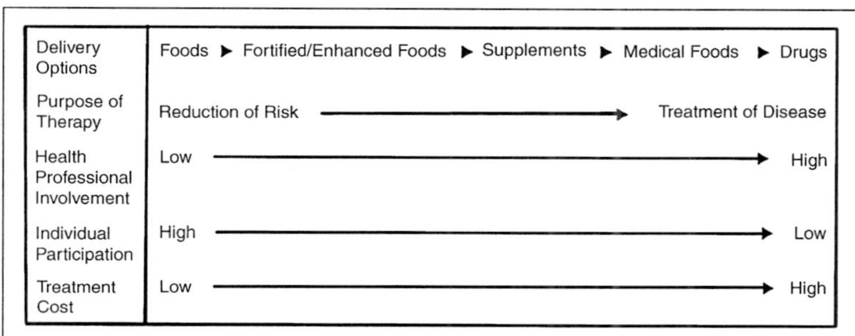

Fig. Role of Functional Foods in Health Care Continuum

On one end of the continuum are public health programmes aimed at reducing disease risk in a large segment of the population through self-directed lifestyle changes. The other end of the continuum is individualized treatment of disease by health care professionals using drugs and other medical interventions. Although the health professional involvement is low in self-directed treatment relative to individualized treatment, an important educational component remains.

New functional foods will continue to expand the continuum, providing additional options for consumers. There is a role for all aspects of this paradigm in our health care system. Functional foods should be integral components of established public health programmes to reduce the risk of specific diseases.

Treatment and prevention of coronary heart disease (CHD) provides an example of this paradigm shift. In the past, recommendations for treating hypercholesterolemia, one of the risk factors for CHD, included dietary and lifestyle interventions along with medication. The dietary and lifestyle interventions included reducing intake of saturated fat and cholesterol, quitting smoking, increasing regular physical activity, and maintaining a healthy body weight. These recommendations, often in conjunction with medication, have been effective strategies for managing heart disease.

The most recent clinical guidelines for treatment of coronary heart disease include therapeutic dietary options for reducing low density lipoproteins (LDL) by consuming specific foods, such as those that contain plant stanols/sterols, increasing intake of soluble fibre, and reducing intake of trans fatty acids. Several food components currently under study may provide additional dietary options in the prevention and treatment of CHD.

TAILORING DIETS FOR SPECIAL NEEDS

Functional foods can address many consumer needs within the new paradigm when used as part of a diet tailored to address the special health needs of a specific group of consumers. In addition to those with needs because of chronic medical conditions, other groups with special needs include women of childbearing age, adolescent girls and boys, athletes, military personnel, and the elderly. For example, improving the health of the elderly in cost effective and consumer-acceptable ways will become even more urgent as the population of individuals 65 years of age and over increases by approximately 50% during the next 27 years.

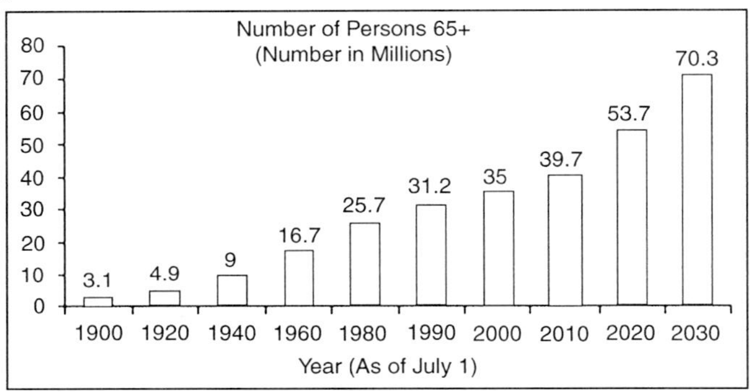

Fig. Projected Increase in Number of Elderly Individuals

The Institute of Medicine reported that poor nutritional status is a major issue for older citizens and that at least four health conditions would benefit from nutritional intervention in either "preventative or treatment modes." Some functional foods are already available for each of these purposes, but more are needed. Many elderly individuals may benefit by expanding their use of functional foods and supplements, particularly where new research can guide their selection of those foods to meet specific needs. It would be unreasonable to expect functional foods to address all of the elderly's medical needs, but functional foods can improve health and wellness, minimize costs, and provide consumers with greater control.

ENCOURAGING THE DEVELOPMENT OF FUNCTIONAL FOODS

As research provides clear evidence of relationships between dietary components and health benefits, the challenge has just begun. Scientific, regulatory, and business frameworks must be in place to evaluate the data for efficacy and safety, ensure effective regulatory oversight, communicate the findings to consumers, and provide incentives that encourage research and development of these novel food products. This report recommends modifications to the existing efficacy and safety evaluation process to ensure a sound scientific underpinning for each proposed functional food, while providing clear information to consumers. Corresponding improvements in the regulatory oversight of new functional components also are proposed. These changes must be implemented now to protect consumer confidence in the safety of the food supply and to encourage the food industry to invest in the development of new functional foods. Science is moving rapidly; industry and government must also move rapidly to ensure that the results are translated into benefits for the consumer.

The functional foods currently available represent only a fraction of the potential opportunities for consumers to manage health through diet. Traditional definitions and arbitrary distinctions between food and medicine should not prevent consumer access to knowledge about the benefits of incorporating functional foods into their diets. Likewise, the framework for providing strong regulatory oversight should not present unnecessary barriers to the development and marketing of functional foods. Where existing terminology and regulatory frameworks are inadequate to address the full scope of benefits and opportunities for functional foods, the terminology and the frameworks must be modified.

Developing a new functional food is an expensive process. Food companies have traditionally funded research for new food product formulations but for functional foods, the stakes are higher—for both food companies and consumers. Government investment in basic and applied research will promote the development of functional foods, but additional incentives are needed to reward private companies that pioneer new health claims.

The research required for a functional food to meet scientific standards for efficacy and safety is a substantial investment, but currently the return on that investment is not exclusive to the company that conducted the research and developed the initial regulatory petition. As soon as the health claim is adequately documented, competing companies can use the claim. Incentives, such as a period of exclusivity or tax incentives, would encourage food companies to pursue functional food development by ensuring a profitable return on successful products.

GENETICALLY MODIFIED ANIMALS

ALLERGENICITY AND HYPERSENSIBILITY

Most food allergens are proteic, well known and identified. As regards the safety of new proteins in foods derived from transgenic animals, the possibility that a new protein could cause an allergic reaction among certain individuals needs to be assessed. FAO has published an approach for assessing the potential allergenicity of new proteins extracted from foods derived from transgenic animals, using the decision tree reproduced in Figure.

If the new protein is produced from a transgene that is known to have allergenic potential, or if the animoacid sequence is similar to that of a known allergenic protein, then the allergenic potential of the new protein can be tested with the serum of individuals with known food allergies. The problem arises for proteins with unknown allergenic potential which cannot be easily assessed, as there is no existing human immunoserum. The problem is further compounded when the new protein derives from an existing protein that has no known allergenic potential.

Thus, assessing the potential allergenicity of a protein from a transgene remains one of the major difficulties in the overall safety assessment of transgenic foods. Assessment of allergenicity requires an understanding of several factors, including the source of the transgene protein, its level of expression, its chemical and physical properties, and its similarities to any known allergen.

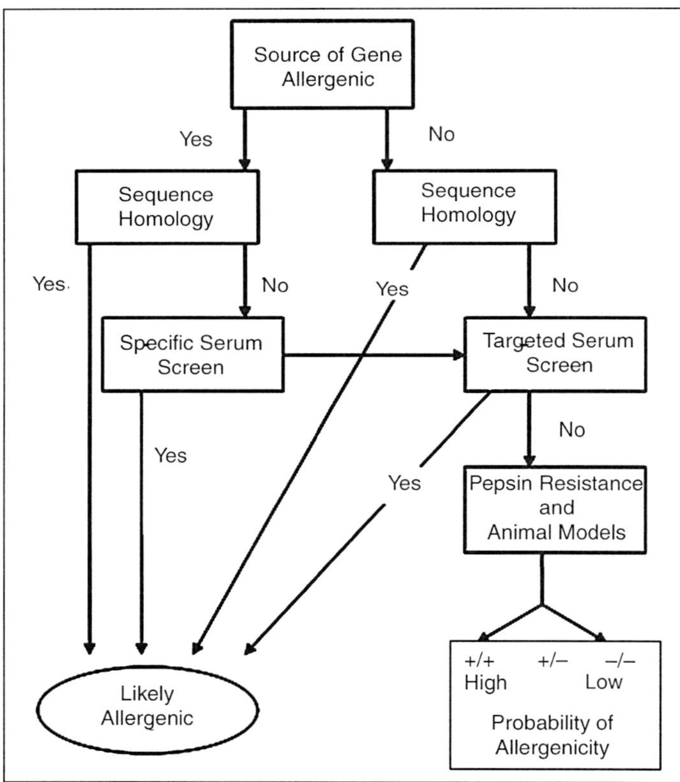

Fig. FAO/WHO 2001, Decision Tree to Assess the Potential Allergenicity of Foods Derived from Biotechnology.

All these factors have to be considered together to have an idea of the potential allergenicity of the new protein. The absence of appropriate and effective methods of analysis is a major problem, as is the feasibility of performing adequate assessments for an increasing number of transgenic products. In this connection, WHO and FAO are working on the development of a standardized protocol for the assessment of allergenicity. The animal model for allergenicity tests of products derived from transgenic animals is recognized as effective. In vivo allergenicity tests among humans would be the best option but raises ethical problems. Post-market surveillance of a product is also advocated, but modalities for doing this are difficult to determine.

Other Bioactive Compounds

Some transgenic animals are designed to express biopharmaceuticals. That these products could retain their bioactivity in the organism after consumption of the food is a safety concern. All proteins and polypeptides are normally broken down in the digestive tract, thus destroying their bioactivity. Only small peptides pass the barrier of the digestive epithelium. The problem remains, however, for people whose digestive epithelium has been compromised by disease or injury. Similarly, the digestive epithelium of young children is permeable to proteins for a period of a few weeks to a few months, during which these proteins could enter the bloodstream. This poses a real problem for bioactive peptides contained the milk. The peptides or proteins could have a negative impact on the digestive system. The production of bactericidal proteins in milk to reduce cattle susceptibility to mastitis from *Staphylococcus aureus* or to control bacterial disease in fish could upset the bacteriological equilibrium of the intestinal flora of consumers.

It could also lead to pathogens resistant to these antimicrobials. Many transgenic fish express a growth hormone gene, which is usually of fish origin. Can this growth hormone cause problems among consumers? The FDA reports that studies have shown that non-human growth hormones, whether of bovine, ovine, porcine or whale origin, are not biologically active among humans. Nor are digested peptide fragments or insulin-like growth factors I secreted in response to GH administration. Moreover, the IGF-I is destroyed by heat along with the active growth hormones. In the (unlikely, given their age) case of young children being fed transgenic fish or milk containing GH, this would be destroyed during heating. The problem however remains for individuals whose digestive tract has been impaired by disease, digestive ulcer or age and which therefore absorbs proteins that have not been broken down and introduces them into the organism. So the risk varies according to the individual. For a susceptible individual, the hazard could have severe consequences, while for a healthy individual of any age, the risk is thought to be low to moderate.

Toxicity

Because proteins are partly broken down in the digestive tract, their direct toxicity is only a mild food safety concern. Moreover, any direct toxicity of proteins would normally be detected during pre-market assessment. The risk is therefore minimal. The problem arises in connection with unintended and unanticipated effects of products derived from GM animals, linked not only to the proteins that are directly expressed by a transgene but also to any unintended by-products. For example, the insertion of a transgene could disturb a specific metabolic process and cause the expression of metabolites with some toxicity. What we need to know is whether there are effective methods of analysis and screening that can detect a toxic substance or a change in metabolism that introduces toxicity. Whole foods can have a tiny concentration of proteins expressed by a transgene or arising as a by-product of the transgene. Such toxicity can be very difficult to detect given the sensitivity limits imposed by laboratory animal numbers, the small quantities of substances tested in whole foods, and the possibility that any negative effect will seldom occur.

Methods include:

- Analysis of the structural/functional relationship of the substance and of its physical and chemical properties and comparison with existing databases of toxic agents to see if the molecule has similarities with a known toxic agent.
- *In vitro method:* Testing the stability of molecules under different conditions or the ability of the molecule to act on other molecules with specific functions. Few *in vitro* tests have been validated and extrapolated *in vivo* as this is difficult, for the method is still fraught with uncertainties.
- *Animal model*: The best indicator of possible toxicity. However, extrapolation remains uncertain, hence the use of safety factors. This model serves to determine acute, subchronic or chronic toxicity. The problem is that the dietary regime needs to be balanced if parallel side effects are not to be caused (high possibility of bias). A precise diet, specific doses and clear protocols are required.

Institutions, Scientific Expertise and International Standards

International agencies have been charged with establishing international standards for GMOs and derived products. Codex is developing principles for risk analysis applied to foods from GMOs. The first principle is to have a pre-market assessment involving case-by-case study of the direct and indirect consequences of inserting a new gene (toxicity, allergenicity, properties of compounds, stability of genome, nutritional effects and other impacts). Codex principles are however for reference only and not binding.

Evaluation of Safety of Foods Derived from Genetically Modified Animals

As concerns examination of the safety of GM foods, the risks associated with genetic engineering are the involuntary modification of the host genetic substance, which can induce changes in food components, including the toxic substances.

Evaluation of the safety of GM foods should therefore include the following procedures:

- Identification of the organism that has been modified and of the source organism(s) of the inserted gene(s). The origin and nature of all the genetic elements that have been introduced into the modified organism need to be identified, including the structural and regulatory sequences and all other parts of the vector sequences. The possibility of horizontal transfer of genes and related consequences also needs to be studied. If the transfer of genes resistant to antibiotics is intended, the possible health implications of concentrations of micro-organisms in the human digestive tract or other parts of the body need to be considered.
- Identification of primary and secondary genetic products, including a description of the characteristics of the inserted gene. Identification of hazards requires an understanding of the expressed inserted genes, the characteristics, concentration and location of expressed food products and the consequences of expression. For each protein resulting from genetic modification, the factors taken into account include possible toxicity and allergenicity from exposure to the expressed protein.
- Evaluation of the safety of new substances anticipated in the foods (*i.e.* proteins, glucides and lipids).
- Assessment of unanticipated effects on the composition of foods, including assessment of changes in the concentration of nutritive or toxic substances occurring naturally, identification of anti-nutritional factor compounds significantly altered in GM foods and assessment of the safety of compounds showing significant modified concentration.
- Evaluation of any toxin produced directly by modification.
- Evaluation of type of treated food, assuming normal manufacturing procedures.
- Evaluation of food consumption patterns, including identification of potential human population consuming GM foods and probable consumption levels, and evaluation of possible consequences if the consumption of modified foods differs from that of equivalent traditional foods.
- Assessment of allergenicity of new foods.

Current approaches to assessing the toxicity of specific proteins are based on internationally approved, acute or chronic testing on laboratory rodents or on fast-growth domesticated species, such as chickens. However, no repeated dose testing with GM foods or food products has yet been published. As regards allergenicity, most scientists now base their evaluations on a decision tree that has been designed to evaluate the potential allergenicity of foods from GM plants, in the light of current knowledge of what is defined as an allergen and how to better identify an allergen.

CONCEPT OF SUBSTANTIAL EQUIVALENCE

The OECD's principal contribution to the evaluation of safety of foods from genetically modified animals has been its publication *Safety evaluation of foods derived by modern biotechnology: concepts and principles*. According to this report, the most practical way of determining the relative safety of a new food product is to determine whether it is substantially equivalent to its conventional counterpart, where this exists. This concept was initially applied to organisms of terrestrial origin but, with certain qualifications, was then also applied to products of aquatic biotechnology. Substantial equivalence recognizes that the purpose of evaluation is not to guarantee absolute safety, but to determine whether the transgenic food presents more risks than its conventional counterpart, should this exist. It is generally agreed that such evaluation requires an integrated, specific and gradual approach. The factors taken into consideration when a transgenic food is compared to its conventional counterpart are (as detailed in the previous chapter on the evaluation of safety of foods derived from transgenic animals):

- Its identity, source and composition; its characteristics in relation to the "natural" product;
- The effects of processing and cooking;
- The processing stage, the dna itself and the products of protein expression of the introduced dna; its nutritional qualities;
- The effects on its function;
- Its potential toxicity and allergenicity, as well as possible secondary effects;
- Consumption of the transgenic food and the nutritional impact of its introduction.

Interactions between environmental factors and animal genotype also need to be taken into account. These variations can influence the phenotype and composition of the modified animal and need to be studied during the evaluation of equivalence. If the food derived from a GMO is judged to be substantially equivalent to its conventional counterpart, it is then considered to present the same level of risk. Otherwise, further tests are needed. Confirmation of substantial equivalence does not mean that the product of a genetically modified animal is identical to the conventional product. Comparison does not take all elements into account and substantial equivalence only provides assurance that the appropriate components of safety of the product will be present in equivalent quantities.

As the comparative approach links the composition of new foods to existing products with a history of safe use, the consequences of introducing new foods into a diet can be foreseen. The concept means that all similarities between foods from genetically modified animals and conventional foods can be judged to be safe. Differences noted during comparison are subject to further meticulous examination which includes the testing of traditional nutritive elements and toxicological or immunological testing or long-term studies, depending on the differences noted. Foods from transgenetic animals might not have any conventional equivalent or the differences between these foods and their conventional equivalents may not be sufficiently characteristic. The history of safe use of the conventional product could not therefore be used to assess the safety of the new food under study.

Substantial equivalence cannot be used to determine the safety of foods for which there is no comparator. However, that a novel food does not bear much resemblance to its counterpart does not necessarily make it dangerous. Its safety needs to be determined by using traditional food safety approaches, with supplementary testing on the consequences of modification.

Post-Approval Monitoring of Foods Derived from Transgenic Animals

It is widely accepted that foods and food products need to be thoroughly evaluated before being placed on the market, if consumers, animal health and the environment are to be protected. If the assessment raises safety concerns, the product will not be approved for commercial release. Some countries therefore fail to see any scientific justification for the post-approval monitoring of these products. Other countries, on the other hand, see the monitoring of possible negative or positive impacts of novel foods as a logical follow-up to the initial science-based risk assessment.

They argue that post-approval monitoring should be mandatory to consolidate the initial pre-market risk assessment, which was based on the then state of scientific knowledge. This could apply to any pre-market assessment. It is recognized that it is sometimes difficult to draw clear conclusions from epidemiological studies, especially when applied to food compounds. Post-approval monitoring is advocated as the only feasible way of demonstrating the absence of possible long-term or unexpected consequences of consuming a novel food. Implementing this strategy is not easy, however.

SCIENCE, INSTITUTIONS, SOCIETY AND POLICIES

There are many scientific uncertainties relating to animal biotechnology. Current understanding is considered to be inexact, uncertain and insufficient as regards technique, methodology and epistemology. There are too many factors, variables and interactions to accurately assess the level of risk. These interactions are not always known or understood. Genetic engineering has unexpected secondary consequences and the aggregate impact of these consequences on intended transgene impact cannot be anticipated. Similarly, genes have pleiotropic effects that are impossible to foresee. It is therefore clearly impossible to envisage all the potential risks. Some of these may not be identified and therefore not studied. On the other hand, analysis of a non-existent risk is a waste of time and resources, and can result in unnecessary costly management measures. In addition, estimating potential risk by using inherently biased methodology must necessarily lead to error. Regulatory bodies make their decisions on the basis of science-based risk assessment.

Scientific uncertainties should therefore be priority research areas in both the private and the public sector. A process of consultation with all stakeholders is required to establish agreed and validated risk assessment protocols and to share results and experiences. There are at least four direct consequences of existing scientific uncertainties: application of the precautionary principle in certain national regulations; extreme precaution in risk analysis; stated need for post-approval monitoring of transgenic animals; and the need for further research.

Policy Context, National and International Agencies

Animal biotechnology raises a number of policy problems. Which standards apply to biotechnology products? A major problem is that the risks concern a variety of areas: human and animal health, animal welfare and the environment. How can these different areas be integrated into the decision-making and regulatory process? The welfare of transgenic animals is usually covered by legislation on animal welfare in general. Is specific regulation on the welfare of transgenic animals required? Another problem is the absence of regulation on transgenic insects.

There is clear need to clarify the scope and limits of different regulatory and control agencies, and national and international organizations as regards the different areas that affected by biotechnology. Should their remit be extended beyond public health and environment to include social and economic aspects, given that these are also relevant? Many different agencies are involved: those responsible for the environment (field releases and trials), food and human health, imports and exports, and animal protection and health. It is important to define individual responsibilities, legitimacy and role. Transgenic animals and derived food products are governed by regulation on novel foods (Novel Food Regulation introduced in 1997 by the European Commission) in Europe and (Novel Food Regulation part of the Food and Drugs Act of October 1999) in Canada. For Canada, the concept of novel food includes any food from plants, animals or micro-organisms that has been genetically modified in such a way that new traits have emerged, or are no longer observed, or are observed at unusual levels. In Canada, protocols are drawn up by Health Canada to assess the safety of novel food. In the USA, the US Federal Drug and Cosmetics Act has been amended to include GM food.

To summarize, approaches differ in many respects:

- *Basic legislation*: Some countries (Canada, USA) have incorporated GMO legislation into existing legislation; others (Australia, Japan) have issued specific legislation for biotechnology, resulting in a mass of sometimes extremely complex regulations.
- *Transparency of decision-making*: Some countries do not consider public opinion (Canada) and only provide a summary of information once a decision has been made, without prior notification. Conversely, others announce their intentions, consult public opinion and issue reports that summarize positions and decisions (EU, USA).
- *Approach to risk assessment and in particular to the principle of precaution*: Some countries seek to eliminate all risk and only decide once every aspect has been assessed; others decide on the basis of existing rather than potential risks.
- *Conceptual approach*: For some, the risk to be assessed concerns the product (Canada), while for others it concerns biotechnology in general (Japan, EU, Australia)
- *Independence of decision-making*: Are all links between governments, scientists, industry and expert committees legitimate? In the USA, there is sometimes a close link between decision-makers and assessors.

Good practices often advocated include greater transparency and public involvement, standardized risk assessment concepts, uniform understanding of the principle of substantial equivalence, and distinction between technique and product. According to the British Better Regulation Task Force, regulation should be transparent, accessible, targeted, complete and proportionate to cost and risk.

Convergence of Policy, Science, Ethics and Public Opinion

Concerns that are not science-based should not guide scientific examination. However, such concerns do inherently influence the scientists' mindset, the way they address questions to policy-makers and their presentation of scientific response to public health and environmental concerns. At the same time, government agencies are influenced in their priorities and decisions by the issues raised by public opinion.

PUBLIC PERCEPTION OF TRANSGENIC ANIMALS AND DERIVED PRODUCTS

The Consumers' Perspective

Recent events regarding the application of modern biotechnology to animals and food production have received considerable attention and raised serious public concern. The introduction of new characteristics into food that differ significantly from those of traditional foods could accentuate these concerns in the future. It therefore seems necessary to keep the public informed of safety assessment procedures applied to novel foods. Another difficulty is the differing consumer needs and expectations in the world regarding food quantity and quality, and the differing nutritional values and cultures. These factors need to be taken into account when examining food production, trade and recent events such as crop enhancement and functional foods. Food safety is a universal concern of consumers who need reassurance that marketed transgenic products have been properly tested for deficiencies and checked for safety. The five eurobarometers conducted in the European Union since 1991 (the last in 2002) give a certain idea of how the European public perceives GMOs.

In many countries this perception is in fact increasingly negative, with concerns relating to:

- The unexpected effects and consequences of gmos on human health and environment;
- The potential irreversibility of negative effects;
- The ethical acceptability of gmos;
- The usefulness of gmos;
- Animal welfare;
- Relational imbalances between producers, farmers and consumers;
- The equality of peoples in the face of biotechnology.

There is a clear distinction in the way GMOs are perceived according to their intended application. On the one hand, we have biomedical applications which are considered useful, morally acceptable and are encouraged, while, on the other, we have agrifood applications which are considered redundant and hazardous to society. Public reaction varies according to familiarity and preoccupation with the subject. However, some observers report a difference between what the public says and what the public does or is. People *think as citizens and not as consumers*. The perception of risk differs between the experts and the public. The way the experts estimate risk does not appear to coincide with the public's perception of risk. There appears to be a lack of trust in regulatory bodies and scientific experts, which in part explains the increasing consultation of public opinion, the greater emphasis on transparency and the release of public information. Assessors have identified three categories of public opinion: unconditional supporters, those who accept the risks and opponents.

Acceptance of GMOs is conditioned by four factors: material values, optimism, confidence and preoccupation with issues linked to GMOs and derived products. Survey responses vary according to connotations underlying related terms, such as "genetic" and "biotechnology", so depend directly on the social and scientific climate at the time of the survey. Consumers believe they have a right to be informed on the products they purchase. However, the question of whether the labelling of transgenic foods is the most appropriate and realistic way of enabling consumers to make enlightened choices is a matter of fierce debate in many countries. The Codex Alimentarius Commission is currently examining this issue.

Several governments have adopted policies and procedures on GMO labelling, and significant differences exist. Provisions for labelling from production to consumption can pose insurmountable problems for countries with limited capacity but anxious to enter international markets. Public opinion in the USA appears to be less animated, and opinions seem stronger on genetically modified animals than on genetically modified crops. The American people appear to be aware of and encourage the potential benefits of biotechnology, with one survey revealing that 62% believe biotechnology will produce additional benefits over the next five years. Such benefits include improved quality, taste and diversity of food, improved human health and nutrition, reduced use of pesticides and insecticides, lower food costs and improved yields. The current use of GM crops does not seem to be a problem for the US population. Do they feel the same way about GM animals? Survey data reveal inadequate methodology, with the use of closed questions and an interpretation of results that needs to be shorn of the deep-seated mental prejudices of both assessors and the public. Questionnaires often focus on GM plants, while it might be illuminating to distinguish between GM plants and GM animals to gauge the level of public concern about genetically modified animals.

Communication between the Public, Assessors and Decision-makers

The OECD countries and intergovernmental organizations are seeking new ways of sharing their experience. They are seeking to boost the dissemination of information and real consumer understanding of the safety issue. Some countries have adopted measures for the public exchange of information on the safety assessment of GM foods. The public are invited to comment on the reports and proceedings of safety assessment bodies.

Some regulatory bodies actively consult the public on food safety and regulations. Some make all information supporting an application available to the public, except confidential commercial information. The Internet is increasingly used to inform the public of safety assessment and approval procedures, and is a good source of information on approved foods and food products.

Some countries are looking into the possibility of using the Internet to make application details accessible to the widest possible audience, in order to make the assessment process as transparent and understood as possible. The OECD's on-line BioTrack site is a useful source of information on regulations in member countries, with information on ministries and agencies, and details of laws, regulations and guidelines. It also has two important databases: one on products that have been marketed; the other on GM field tests carried out in OECD countries. Other approaches adopted with huge success in selected countries include the appointment of consumer representatives to safety assessment committees. Consensus conferences have also been attempted with varying degrees of success.

Ethical Considerations

Up until now, the idea of food products from transgenic animals has not been well received by consumers. Surveys have constantly shown that the public is better disposed towards transgenic plants than transgenic animals. There is greater reluctance towards trials involving animals and changes in animal make-up and therefore broader possible implications. Various cultures and religions limit or forbid the consumption of certain foods of animal origin. The public seems however to be more disposed to consume by ingestion or injection pharmaceutical products derived from transgenic animals. Various ethical issues have been raised.

Distribution of Benefits

One preoccupation is the possible impact of biotechnology on agricultural structure, especially the potential elimination of small livestock farms in the face of the cost of producing transgenic animals. People are worried about the impact of a dismantled agricultural society on the environment, sustainable agriculture and quality of life. The public points to the concentration of production and its direct consequences on markets and consumers, and is wary of the dependence of livestock farmers and food producers on multinationals. There is mental conflict between the law of the market and the traditions, cultural identities and needs of society.

Nutrition and Food Safety

There seems to be general consensus among scientists that the consumption of food derived from transgenic animals is no more hazardous than the consumption of conventional products. However, the public thinks differently. Some people refer to specific diets based on cultural or religious considerations. For example, the consumption of an animal or a food from a transgenic animal that contains proteins or genes of animals whose consumption is forbidden by culture or religion could pose a problem. Another point often stressed is the social divide from different levels of access to biotechnology in developing and developed countries.

Respect for Life and the Unnatural Nature of GMOs

Some people do not believe that technical and economic pressures justify the commercialization of life and the patenting of living organisms. Overstepping of the species barrier and the large-scale industrial production of genetic animals are seen as contrary to nature. This argument is frequently refuted by biologists who stress that the continuous natural evolution or the human selection of animal genotypes has favoured certain genes and eliminated others. Finally, biotechnology and conventional livestock practices might seem to pose similar problems, in that both seek maximized productivity and both cause animal dysfunctioning and welfare and health problems. However, such problems seem to be magnified (some would say disproportionately) in the context of biotechnology.

Animal biotechnology is a recent science and therefore difficult to view in the light of experience. Many uncertainties and many doubts exist. Information currently available will have to be constantly updated with data and results of analyses and trials of transgenic animals and derived food products, so that techniques, regulations and public opinion can evolve and adapt; so that doubts can be clarified and problems resolved; and so that agreed protocols of analysis and international standards can be determined. Given the rapid progress being made and the prospects of animal biotechnology, it is almost certain that improved techniques will boost benefits. However, the trade-off between risk, cost and benefit will always have to be examined. Ongoing oversight, greater transparency of results from scientific trials, greater transparency of regulatory and decision-making processes and closer cooperation among all stakeholders are required for the safety evaluation of products derived from transgenic animals.

HEREDITARY DNA OF TRANSGENIC ANIMAL

A transgenic animal is an animal whose hereditary DNA has been augmented by addition of DNA from a source other than parental germplasm through recombinant DNA techniques. Transfer of genes or gene constructs allows for the manipulation of individual genes rather than entire genomes. There has been dramatic advances in gene transfer technology in the last two decades since the first successful transfer was carried out in mice in 1980. The technique has now become routine in the mouse and resulting transgenic mice are able to transmit their transgenes to their offspring thereby allowing a large number of transgenic animals to be produced. Successful production of transgenic livestock has been reported for pigs, sheep, rabbits and cattle. The majority of gene transfer studies in livestock have, however, been carried out in the pig. Although transgenic cattle and sheep have been successfully produced, the procedure is still inefficient in these species.

Transgenesis offers considerable opportunity for advances in medicine and agriculture. In livestock, the ability to insert new genes for such economically important characteristics as fecundity, resistance to or tolerance of other environmental stresses would represent a major breakthrough in the breeding of commercially superior stock. Another opportunity that transgenic technology could provide is in the production of medically important proteins such as insulin and clotting factors in the milk of domestic livestock. The genes coding for these proteins have been identified and the human factor IX construct has been successfully introduced into sheep and expression achieved in sheep milk. Moreover, the founder animal has been shown to be able to transmit the trait to its offspring. To date, the majority of genes transferred into sheep have been growth hormone encoding gene constructs. Unfortunately, in most cases the elevated growth hormone levels have resulted into a clinical diabetes situation leading to an early death of the transgenic sheep. Transgenic sheep have recently been generated which express the visna virus envelope gene.

The first reports of the production of transgenic animals created a lot of excitement among biological scientists. In the field of animal breeding, there were diverse opinions on how the technology might affect livestock genetic improvement programmes. Some believed that it would result in total reorganisation of conventional animal breeding theory while others considered the technology as an extension of current animal breeding procedures which, by broadening the gene pool, would make new and novel genotypes available for selection. Application of the technology in animal improvement is still far from being achieved. However, consideration needs to be given to its potential role in this field. Smith *et al*. presented a comprehensive evaluation of strategies for developing, testing, breeding and disseminating transgenic livestock in the context of quantitative improvement of economic traits.

An important contribution of transgenic technology is in the area of basic research to study the role of genes in the control of physiological processes. The understanding of the molecular control of life processes has important implications for both medicine and agriculture. For example, the generation (through mutation of an endogenous gene) of an organism which lacks a specific gene is a powerful tool to investigate the function of the gene product. This type of genetic analysis has been facilitated by the availability of in vitro cultures of embryonic stem cells from mice. Recent advances in *in vitro* technology (*in vitro* fertilisation and maturation) will increase the number of zygotes available for gene transfer purposes. This, plus the utilisation of embryonic stem cell and primodial germ cell technologies should enhance the efficiency of gene transfer in cattle and sheep considerably.

GENETIC CHARACTERISATION OF ANIMAL GENETIC RESOURCES

Animal Genetic Resources Division is engaged in physical characterization through field survey, sustainable utilization and conservation of indigenous livestock and poultry breeds. Systematic field surveys have been undertaken to assess the phenotypic traits, population status of the breeds and potential population groups, besides the socio-economic conditions of the farmers and communities. In the same vein, the production performance of local breeds and their interaction with local ecology have been evaluated and recorded. Based on the information, new strategies have been formulated for improvement and conservation of the breeds in the farmers' conditions. The in situ conservation has been implemented for different species. In addition, the Division is also working in the frontier areas of ex-situ conservation of germplasm for long term preservation.

Developing countries are endowed with the majority of the global domestic animal diversity-landraces, strains or breeds. Some livestock breeds in these countries are in immediate danger of loss through indiscriminate crossbreeding with exotic breeds. The importance of indigenous livestock breeds lies in their adaptation to local biotic and abiotic stresses and to traditional husbandry systems. However, most of these animal genetic resources are still not characterised and boundaries between distinct populations are unclear. In such cases breeds are defined on the basis of subjective data and information obtained from local communities. Reliance on these criteria as the basis for classification for utilisation and/or conservation may be misleading. Additionally, historical evidence is not always accurate, relying as it often does on subjective judgements. Archival research can reveal much about the original type of a breed or strain but it is molecular genetic evidence which is factual and precise. It is in this sphere that biotechnology has an important role.

Genetic uniqueness of populations is measured by the relative genetic distances of such populations from each other. Polymorphism in gene products such as enzymes, blood group systems and leukocyte antigens which have traditionally been used for measuring genetic distance are being rapidly replaced by polymorphism at the level of DNA, both nuclear and mitochondrial as a source of information for the estimation of genetic distances. The first DNA polymorphism to be used widely for genome characterisation and analysis were the restriction fragment length polymorphism (RFLP) which detect variations ranging from gross rearrangements to single base changes.

Minisatellites sequences of 60 or so bases repeated many hundreds or thousands of times at one unique locus within the genome have been used to generate DNA fingerprints typical of individuals within species. Microsatellites repeats of simple sequences, the commonest being dinucleotide repeats are abundant in genomes of all higher organisms, including livestock. Polymorphism of microsatellites takes the form of variation in the number of repeats at any given locus and is generally revealed as fragment length variation in the products of polymerase chain reaction (PCR) amplification of genomic DNA using primers flanking the chosen repeat sequence and specific for a given locus. Ease of identification and of sequence determination and need for only small amounts of DNA, are some of the advantages of microsatellites. Additionally, because microsatellite polymorphism can be described numerically, they lend themselves to computerised data handling and analyses. Microsatellites can be used in non-PCR systems in a way similar to minisatellite probes.

Randomly amplified polymorphic DNA (RAPD) has been extensively used for genetic characterisation of a wide range of organisms. The technique uses short (up to 10 bases) primers to amplify nuclear DNA in the PCR. The procedure does not require knowledge of the sequence of DNA under study; primers are designed randomly. The basis of the polymorphism detected by this method is that products are either generated in PCR or not.

Complete sequencing of the genome is the ultimate form of genetic characterisation. Sequencing has traditionally been expensive and laborious, but with the advent of automated sequencing this is changing rapidly. However, sequencing is unlikely to be used as a technique of choice for genetic characterisation.

Nei and Takezaki reviewed statistical methods for estimating genetic distances and for constructing phylogenetic trees from DNA sequence data and concluded that different analytical methods may produce different results. Teale *et al.* commented on what considerations to be made in using DNA polymorphism data for genetic distance estimation and cautioned that great care has to be taken in selecting characterisation methods and in interpreting the resulting data. While recognising the importance of the uniparental mode of inheritance of mitochondrial DNA in detecting underlying population structure not discernible from analyses of nuclear DNA, Loftus *et al.* concluded that mitochondrial DNA analysis may not be sufficient to resolve breed differences within Africa. MacHugh *et al.* suggested that microsatellite polymorphism may be more suitable when trying to discriminate between closely related populations. Regardless of which method is used, the ultimate goal in genetic characterisation for conservation is to obtain a measure of available diversity.

CONSERVATION OF ANIMAL GENETIC RESOURCES

The terms conservation, preservation, ex situ and in situ are used here according to the definition given by FAO. There are several ways, differing in efficiency, technical feasibility and costs, to conserve animal genetic resources. Developing and utilising a genetic resource is considered the most rational conservation strategy. However, there are cases where ex-situ approaches are the only alternatives. *Ex-situ* approaches include: maintenance of small populations in domestic animal zoos; cryopreservation of semen (and ova); cryopreservation of embryos; and some combinations of these. Brem *et al.* reviewed biotechnologies for *ex-situ* conservation.

Cryopreservation of gametes, embryos or DNA segments can be quite an effective and safe approach for breeds or strains whose populations are too small to be conserved by any other means. The safety of these methods has been demonstrated by background irradiation studies. For example, studies based on irradiation of mouse embryos exposed to the equivalent of hundreds of years of background mutation showed no detectable damage. Regeneration of offspring following transfer of frozen-thawed embryos has been successful for all major domestic species, except the buffalo. In cattle, the transfer of frozen-thawed embryos is now a commercial practice and embryo survival rate after thawing can be as high as 80% with a pregnancy rate of about 50%. Cryopreservation of oocytes followed by successful fertilisation and live births have been achieved in the mouse. Cryopreserved bovine oocytes have been successfully matured and fertilised in vitro and zygotes developed to blastocyst stage. These trends strongly suggest that long-term cryopreservation of mammalian oocytes is possible.

Respective pregnancy rates of 58 and 50% for fresh and frozen-thawed *in vitro* produced embryos have been reported. Also, calves have been produced from transfer of both split and frozen-thawed *in vitro* produced embryos.

Economic aspects of genetic conservation in farm animals has been assessed by Brem *et al*.. The study concluded that costs of *ex situ* live animal conservation was moderate to high while costs of long-term cryopreservation of gametes were low.

Development in genetic engineering, cryobiology, cell biology and embryology will provide techniques that may enhance our ability to preserve germplasm *in vitro*. Techniques such as transfer of DNA within and between species and the production of viable transgenic animals are far from practical application. However, biotechnology will certainly contribute newer and cheaper methods for preservation such as storage of catalogued DNA. At present, other than live animal and embryo preservation, the other techniques do not allow preservation of genomes in a form which can be reactivated *in toto* at a later stage, but they permit the preservation of individual genes or gene combinations for possible future regeneration.

Conservation of indigenous animal genetic resources should be one of the priority livestock development activities for developing countries. The critical importance of these resources to their owners in developing countries need not be emphasised. Their importance to developed countries is also becoming evident as indicated by the increasing importation of tropical germplasm by these countries. It is highly likely that these resources will become of increasing importance to the industrialised countries either as sources of unique genes or when environmental concerns necessitate change in production systems. Developed countries should, thus, assist in the conservation and development of these resources. Technology for cryopreservation of semen and embryo is sufficiently developed to be applied in developing countries. What is missing is financial support to implement conservation programmes. Such support has been provided for world-wide conservation activities for plant germplasm. There is also a strong case for support of animal genetic resources conservation.

STRUCTURE/FUNCTION CLAIMS FOR CONVENTIONAL FOODS

The FDC Act provides that products that are "intended to affect the structure or any function of the body" generally are subject to regulation as drugs, but this does not apply in the case of food. It has long been recognized that a food may make labeling representations about its dietary impact on the structure or function of the human body, provided that the particular claim used does not also represent that the food will cure, mitigate, treat, or prevent disease and provided further that the claim does not trigger some other requirement for FDA preclearance.

In practice, companies have made a few claims of this type that FDA generally has accepted over the years, without asserting that the claim creates drug status or that the claim is a health claim that requires compliance with health claim requirements. For example, claims of the general type "calcium helps build strong bones" or "protein helps build strong muscles" have long been made in food labeling and appear generally to have been accepted by FDA as appropriate claims about the impact of a food on the structure or function of the body. In principle, it would appear that this type of claim could be extended.

For example, it would appear to be proper to make a truthful and nonmisleading claim to the effect that a substance in a food "helps maintain a normal, healthy cardiovascular system" without triggering either drug status or requirements for approval of a health claim. However, there is considerable uncertainty about how far this type of structure/function claim can be "pushed" before FDA will assert either drug status or health claim status.

In a preamble in the Federal Register of Sept 23, 1997, FDA stated as follows:

- FDA points out that the claim that cranberry juice cocktail prevents the recurrence of urinary tract infections...is a claim that brings the product within the "drug" definition...because it is a claim that the product will prevent disease. However, a claim that cranberry products help to maintain urinary tract health may be permissible on...cranberry products in conventional food form...if it is truthful, not misleading, and derives from the nutritional value of cranberries. If the claim derives from the nutritive value of cranberries, the claim would describe an effect of a food on the structure or function of the body and thus fall under one exception to the definition for the term "drug"....The claim is not a health claim because no disease is mentioned explicitly or implicitly...

Clearly, there is considerable opportunity to make labeling claims about the favourable impact of a food on the normal, healthy structure or function of the human body. However, some have maintained that FDA's insistence on derivation from "nutritional" or "nutritive" value is not a correct statement of the law.

As defined by the FDC Act, the term "drug" means "...articles intended to affect the structure or any function of the body of man," and "food" includes:

- Articles used for food or drink for man or other animals,
- Chewing gum, and
- Articles used for components of any such article.

In reviewing the definition, the U.S. Court of Appeals for the Seventh Circuit stated:

- When the statute defines "food" as "articles used for food," it means that the statutory definition of "food" includes articles used by people in the ordinary way most people use food—primarily for taste, aroma, or nutritive value. To hold...that articles used as food are articles used solely for taste, aroma or nutritive value is unduly restrictive since some products such as coffee or prune juice are undoubtedly food but may be consumed on occasion for reasons other than taste, aroma, or nutritive value.

This interpretation has been accepted by other federal courts. Thus, the courts have recognized that the food exemption from the drug definition in the FDC Act is not limited to nutritional or nutritive substances. Established case law, an article may be a food within the meaning of the FDC Act if it is used "primarily" for taste, or for aroma, or for nutritional value; in addition, sometimes a food—such as coffee or prune juice—will not even be used for any of these three purposes.

The exclusion from "drug" status for a "food" in the FDC Act is therefore not properly limited only to products that are "nutritional" or "nutritive"—because "food" is much broader than that. Since a food's effects need not be of a nutritional nature, there is no apparent reason why a food may not properly provide labeling information about its effects on the structure or function of the body that do not derive from nutritional value.

Indeed, in American Health Products v. Hayes, the U. S. District Court for the Southern District of New York stated plainly:

- ...if an article affects bodily structure or function by way of its consumption as a food, the parenthetical precludes its regulation as a drug notwithstanding a manufacturer's representations as to physiological effect .... The presence of the parenthetical in suggests that Congress did not want to inhibit the dissemination of useful information concerning a food's physiological properties by subjecting foods to drug regulation on the basis of representations in this regard.

Thus, the courts have recognized that coffee may be used to help stay alert, or that prune juice may be used to help promote regularity, and that labeling claims about this type of physiological effect are appropriate for a food and do not create drug status—regardless of whether such effects and claims derive from the nutritional/nutritive value of the food. Even if it were true that a structure/function claim for a food should derive from nutritional value, the Agency's statements about the meaning of the term have been inconsistent.

In the same 1997 *Federal Register* document, FDA stated that even though the term "statement of nutritional support" was used by Congress, FDA chose not to use the term in the regulations "because many of the substances that can be the subject of this type of claim do not have nutritional value. Thus, the term 'statement of nutritional support' is not accurate in all instances."

One could argue that the FDA objection is in conflict with the express intention of Congress to give a broad meaning to "nutritional, " and therefore contrary to law. Nevertheless, this FDA statement certainly suggests an FDA view that nutritional is a concept that should be interpreted critically and narrowly. However, in the context of defining the scope of a nutrient content claim, FDA proceeded in the opposite direction and asserted that the term "nutrient" is not narrow at all, but instead very broad and includes many substances that traditional nutritionists might not regard as nutritional. In this context, FDA stated that "nutrient" encompasses a long list of examples included in a discussion between Senators Metzenbaum and Symms before passage of NLEA in 1990.

The quoted list of agreed-upon examples of nutritional substances includes:

- Primrose oil, black currant seed oil, cold pressed flax seed oil, "Barleygreen" and similar nutritional powdered drink mixes, Coenzyme Q10, enzymes such as:
  - Bromelain and quercetin,
  - Amino acids,
  - Pollens,
  - Propolis,
  - Royal jelly,
  - Garlic,
  - Orotates,
  - Calcium-EAP,
  - Glandulars,
  - Hydrogen peroxide,
  - Nutritional antioxidants such as superoxide dismutase, and Herbal tinctures.

Moreover, both of these discussions fail to reference the Agency's own definition of nutritive value:

- "A value in sustaining human existence by such processes as promoting growth, replacing loss of essential nutrients that cannot be produced in sufficient quantities by the body, or providing energy."

Considering the Congressional intent and some of the Agency's own statements, it would appear that even if FDA were correct in tying structure/

function claims to nutritional value, the meaning of nutritional in this context would need to be regarded very broadly. As stated in the summary report of a public meeting on the conceptual framework for structure/function claims for conventional foods posted on FDA's website, "Nutritive value cannot be defined simply in terms of source, dose or biochemical composition."

CLAIMS ABOUT SPECIAL DIETARY USES

Since 1938 the FDC Act has recognized that it is proper for a food to be labeled with claims "for special dietary uses." FDA is given authority to issue regulations that require additional informative labeling for foods that are represented for special dietary uses. In the past, FDA issued regulations requiring certain additional labeling information for certain types of foods for special dietary uses. There continues to be a regulation of this type that governs the use of "hypoallergenic" labeling. This regulation provides that if a food is represented "for special dietary use by reason of the decrease or absence of any allergenic property or by reason of being offered as food suitable as a substitute for another food having an allergenic property," the label of the food must bear certain information, including the "quantity or proportion of each ingredient." FDA has said that if a claim that otherwise would require FDA approval as a health claim is already authorized by a regulation concerning special dietary use, FDA will not require that a new health claim regulation also be issued. If a company is interested in using a new labeling claim that would fall within the definition of a health claim, then instead of petitioning FDA to issue an approving health claim regulation, the company may be able to petition the Agency to issue a special dietary use labeling regulation. However, this is a largely theoretical option. In practice, FDA has avoided issuing new special dietary use regulations in recent years; indeed, the Agency has been revoking some of these regulations.

GENERAL FREEDOM TO USE STATEMENTS THAT ARE NOT 'FALSE OR MISLEADING IN ANY PARTICULAR'

In addition to the various authorizations to use particular types of health-related claims, it should also be remembered that the FDC Act contains no general requirement that statements included in labeling of FDAregulated foods must be approved by FDA prior to use. Instead, requirements for FDA preclearance are confined to certain specific types of labeling statements and except for such specific requirements, food labeling generally may include any statement, so long as it is truthful and not misleading in any particular.

SCIENTIFIC STANDARDS FOR EVALUATING A PROPOSED CLAIM

The evidence supporting a functional food claim must meet certain standards. The level of support for these claims ranges from significant scientific agreement (SSA) for approved health claims to "FDA has determined that this evidence is limited and not conclusive" and two other qualifying levels of data within this range for qualified health claims to "competent and reliable scientific evidence" for structure/function claims. The application of any standard is intended to be objective and based on a body of sound and relevant scientific data. It is also intended to be flexible, recognizing the variability in the amount and type of data needed to support the validity of different substance/health relationships.

SIGNIFICANT SCIENTIFIC AGREEMENT

When FDA evaluates a petition for approval of a health claim, it issues a regulation only when it determines that there is "significant scientific agreement" that the claim is supported by scientific evidence. This evaluation considers whether experts would agree that the claim is valid based on the totality of publicly available scientific evidence.

In explaining its SSA standard for health claims, FDA stated:

- The standard of scientific validity for a health claim includes two components:
  - That the totality of the publicly available evidence supports the substance/disease relationship that is the subject of the claim, and
  - That there is SSA among qualified experts that the relationship is valid.

FDA further described SSA:

- FDA's determination of when SSA has been achieved represents the Agency's best judgment as to whether qualified experts would likely agree that the scientific evidence supports the substance/disease relationship that is the subject of a proposed health claim. The SSA standard is intended to be a strong standard that provides a high level of confidence in the validity of a substance/disease relationship. SSA means that the validity of the relationship is not likely to be reversed by new and evolving science, although the exact nature of the relationship may need to be refined. Application of the SSA standard is intended to be objective, in relying upon a body of sound and relevant scientific data; flexible, in recognizing the variability in the amount and type of data needed to support the validity of different substance/disease relationships; and responsive, in recognizing the need to re-evaluate data over time as research questions and experimental approaches are refined. SSA does not require a consensus or agreement based on unanimous and incontrovertible scientific opinion. However, on the continuum of scientific discovery that extends from emerging evidence to consensus, it represents an area on the continuum that lies closer to the latter than to the former.

FDA has specifically mentioned that SSA is not consensus:

- Although SSA is not consensus in the sense of unanimity, it represents considerably more than an initial body of emerging evidence. Because each situation may differ with the nature of the claimed substance/disease relationship, it is necessary to consider both the extent of agreement and the nature of the disagreement on a case-by-case basis. If scientific agreement were to be assessed under arbitrary quantitative or rigidly defined criteria, the resulting inflexibility could cause some valid claims to be disallowed where the disagreement, while present, is not persuasive.

In assessing the validity of codified health claims, FDA has considered three types of evidence:

- *Epidemiology*: Data derived from observational studies assessing associations between food substances and disease;
- *Biological Mechanisms:* Data derived from chemical, cellular, or animal models investigating plausible mechanisms of action for food substances;
- *Intervention Trials*: Controlled assessment of clinical food substance interventions in the human population. The "gold standard" is the randomized controlled clinical trial. FDA felt that these combinations of data met the SSA standard of proof. A number of sequential threshold questions are addressed in the review of the scientific evidence:
- Have studies appropriately specified and measured the substance that is the subject of the claim?
- Have studies appropriately specified and measured the disease that is subject of the claim?
- Are all conclusions about the relationship between the substance and the disease based on the totality of the publicly available scientific evidence?

The assessment of SSA then derives from the conclusion that a sufficient body of sound, relevant scientific evidence shows consistency across different studies and among different researchers and permits the key determination of whether a change in the dietary intake of the substances will result in a change in a disease or structure/function endpoint.

WEIGHT OF THE SCIENTIFIC EVIDENCE

In its December 2002 announcement regarding qualified health claims, FDA indicated that codified health claims would still require substantiation meeting the SSA standard. In its initial guidance on qualified health claims, the Agency said it would use a "weight of the scientific evidence" standard to establish qualified health claims.

At that time, the following was proposed:

- To meet the criteria for a qualified health claim, the petitioner would need to provide a credible body of scientific data supporting the claim. Although this body of data need not rise to the level of SSA defined in FDA's previous guidance, the petitioner would need to demonstrate, based on a fair review by scientific experts of the totality of information available, that the "weight of the scientific evidence" supports the proposed claim. The test is not whether the claim is supported numerically, but rather whether the pertinent data and information presented in those studies is sufficiently scientifically persuasive. For a claim that meets the WOSE standard, the Agency would decline to initiate regulatory action, provided the claim is qualified by appropriate language so consumers are not misled as to the degree of scientific uncertainty that would still exist.
- FDA anticipates that this policy will facilitate the provision to consumers of additional, scientifically supported health information. FDA expects that, as scientific enquiry into the role of dietary factors in health proceeds, particular qualified health claims will be further substantiated, while for other qualified health claims the "weight of the scientific evidence" will shift from "more for" to "more against." It is conceivable, therefore, that the information provided to consumers through qualified health claims in food labeling could change over time. FDA nevertheless believes that the dissemination of current scientific information concerning the health benefits of conventional foods and dietary supplements should be encouraged, to enable consumers to make informed dietary choices yielding potentially significant health benefits.

In July 2003, FDA published *Guidance for Industry and FDA for Interim Evidence-based Ranking System for Scientific Data*. As stated in this document, "FDA has tentatively chosen to model its evidence-based rating system on that of the Institute for Clinical Systems Improvement as adapted by the American Dietetic Association (ADA) with modifications specific to FDA. In making this tentative decision, FDA relied on criteria for evaluating evidence-based rating systems as reviewed and critiqued by the Agency for Healthcare Research and Quality. FDA also found the modifications from ADA to be particularly useful as they considered diet and health relationships, whereas other groups focused on drug and treatment applications."

The elements of the evidence-based rating system include:

- Define the substance/disease relationship;
- Collect and submit all relevant studies;
- Classify, and therefore rate, each study as to type of study;
- Rate each study for quality;

- Rate the strength of the total body of evidence; and
- Report the "rank."

The criteria used to determine the ranking of scientific evidence would include: satisfying the necessary quality level for studies, meeting prescribed design types, considering the number of individuals tested, and confirming that study results are relevant to the target population. When rating the strength of the total body of evidence, "the rating system is based on three factors: quantity, consistency, and relevance to disease risk reduction in the general population or target subgroup." The first level of ranking meets the SSA standard and reflects "a high level of comfort" that the claimed substance/disease relationship is scientifically valid. The second level is the highest level for a qualified health claim and represents "a moderate/good level of comfort" that the claimed relationship is scientifically valid. Qualified experts would rank the relationship as "promising," but not definitive. The third level represents "a low level of comfort" that the claimed relationship is scientifically valid.

The fourth level is the lowest level for a qualified health claim and represents "an extremely low level of comfort" that the claimed relationship is scientifically valid. "If the scientific evidence to support the substance/disease relationship is below that described as the fourth level, no claim will be appropriate," FDA stated. Shortly after publication of FDA's guidance on WOSE, the U.S. District Court for the District of Columbia ruled in *Whitaker v. Thompson* that "credible evidence" rather than "weight of the evidence" is the appropriate standard for FDA to apply in evaluating qualified health claims.

Thus, FDA's evaluation of WOSE will be tempered by the test of "credible evidence". The IFT Expert Panel believes the guidance can serve as a useful tool and assist in evaluating data. However, in the final analysis, most decisions will be based more on subjective judgment than on quantitative analysis. Therefore, a WOSE standard, tempered by the test of "credible evidence, " should be the basis for qualified health claims.

COMPETENT AND RELIABLE SCIENTIFIC EVIDENCE

In November 2004, FDA provided guidance to industry for determining whether the available information constitutes "competent and reliable scientific evidence" for structure/function claims for dietary supplements, including:

- Does each study or piece of evidence bear a relationship to the specific claim?
- What are the individual study's or evidence's strengths and weaknesses?
- If multiple studies exist, do the studies that have the most reliable methodologies suggest a particular outcome?
- If multiple studies exist, what do most studies suggest or find/does the totality of the evidence agree with the claim?

6

Embryonic Development of Animals

After cleavage, the dividing cells, or morula, becomes a hollow ball, or blastula, which develops a hole or pore at one end.

BILATERANS

In bilateral animals, the blastula develops in one of two ways that divides the whole animal kingdom into two halves. If in the blastula the first pore (blastopore) becomes the mouth of the animal, it is aprotostome; if the first pore becomes the anus then it is a deuterostome. Theprotostomes include most invertebrate animals, such as insects, worms and molluscs, while the deuterostomes include the vertebrates. In due course, the blastula changes into a more differentiated structure called the gastrula.

The gastrula with its blastopore soon develops three distinct layers of cells (the germ layers) from which all the bodily organs and tissues then develop:

- The innermost layer, or endoderm, gives rise to the digestive organs, the gills, lungs or swim bladder if present, and kidneys or nephrites.
- The middle layer, or mesoderm, gives rise to the muscles, skeleton if any, and blood system.
- The outer layer of cells, or ectoderm, gives rise to the nervous system, including the brain, and skin or carapace and hair, bristles, or scales.

Embryos in many species often appear similar to one another in early developmental stages. The reason for this similarity is because species have a shared evolutionary history. These similarities among species are called homologous structures, which are structures that have the same or similar function and mechanism, having evolved from a common ancestor.

HUMANS

Humans are bilaterans and deuterostomes—the anus develops first. In humans, the term embryo refers to the ball of dividing cells from the moment the zygote implants itself in the uterus wall until the end of the eighth week after conception. Beyond the eighth week after conception (tenth week of pregnancy), the developing human is then called a fetus.

HISTORY

As recently as the 18th century, the prevailing notion in human embryology was preformation: the idea that semen contains an embryo — a preformed, miniature infant, or "*homunculus*" — that simply becomes larger during development. The competing explanation of embryonic development was *epigenesis*, originally proposed 2,000 years earlier by Aristotle. According to epigenesis, the form of an animal emerges gradually from a relatively formless egg. As microscopy improved during the 19th century, biologists could see that embryos took shape in a series of progressive steps, and epigenesis displaced preformation as the favoured explanation among embryologists. Modern embryological pioneers include Gavin de Beer, Charles Darwin, Ernst Haeckel, J.B.S. Haldane, and Joseph Needham, while much early embryology came from the work of Aristotle and the great Italian anatomists: Aldrovandi, Aranzio, Leonardo da Vinci, Marcello Malpighi, Gabriele Falloppia, Girolamo Cardano, Emilio Parisano, Fortunio Liceti, Stefano Lorenzini, Spallanzani, Enrico Sertoli, Mauro Rusconi, etc. Other important contributors include William Harvey, Kaspar Friedrich Wolff, Heinz Christian Pander, Karl Ernst von Baer, and August Weismann.

After the 1950s, with the DNA helical structure being unravelled and the increasing knowledge in the field of molecular biology, developmental biology emerged as a field of study which attempts to correlate the genes with morphological change, and so tries to determine which genes are responsible for each morphological change that takes place in an embryo, and how these genes are regulated.

MODERN EMBRYOLOGY RESEARCH

Currently, embryology has become an important research area for studying the genetic control of the development process (*e.g.* morphogens), its link to cell signalling, its importance for the study of certain diseases and mutations and in links to stem cell research.

HISTORICAL REVIEW OF EMBRYOLOGY

Developmental biology has a long history, and its is useful to look into the past in order to lay a foundation for understanding the field as it exists today. Although we know nothing of the earliest, prehistoric speculations, the thoughts and concepts of *Aristotle* and other early philosopher-scientists were written down during the *"Golden Age"* of ancient Greece. Important among the surviving treatises of that time is *De Generatione Animalium* by *Aristotle,* who lived from 384 to 322 B. C. In this treatise, *Aristotle* questioned (among other things) whether the embryo is an already performed individual that merely unfolds and grows during development, or whether it progressively differentiates from a formless beginning into a unique, complex, adult individual. *Aristotle* clearly favoured the second view, which we now call *epigenesis;* non-etheless, others held to the first view, which we now call *preformation.* This controversy remained unresolved for some 2,000 years, not being put to rest until early in the 19th century.

During the 17th century, a time of rebirth of scientific enquiry, nearly all biologists favoured the preformation theory. Writing in 1669, *Jan Swammerdam* reported his observation that, in insects, miniature larvae are present in the "*egg*" and that gradual changes could be observed before a butterfly emerged. Actually, his "*egg*" was the pupa, or cocoon—a far cry from the "*real*" egg. In 1694, *Antonvan Leeuwenhoek,* an early Dutch microscopist, first observed living sperm. Until then, the semen was thought merely to stimulate or nourish the egg. With this recognition that living cells were present in semen, two versions of the preformation concept developed: (1) the egg contained the preformed miniature individual, and (2) the sperm did. Thus, "*ovists*" and "*spermists*" sought evidence to support their respective views. In 1964, *Nicklaas Hartsoeker* drew a figure of a miniature human (*homunculus*) inside a sperm, presumably representing what he saw through a microscope. (Like the "*canals*" on Mars, observations such as this demonstrate that we se what we look for, not what we look at.) By 1745, however, *Charles Bonnet* had demonstrated conclusively that unfertilized eggs sometimes develop into normal individuals (parthenogenesis). This, of course, was an exceptionally strong argument that the egg, not the sperm, was the bearer of the preformed individual.

Despite these preformationist views, epigenesis had its adherents, too. As early as 1651, *William Harvey,* who is famous for demonstrating the continuous circulation of the blood, considered the egg to be a formless, undifferentiated mass from which the embryo developed. He is thought to have coined the term "*epigenesis*", following *Aristotle.* Despite the authority of *Aristotle* and *Harvey,* no other 17th century biologist of note supported the concepts of epigenesis. More than 100 years later, *Caspter Friedrich Wolff* (1738-1794) observed microscopically the successive stages of development of the chick egg. However, he found no evidence of a preformed chick in the egg, only progressive growth and gradual development from a simple to a complex form. The preformation theory received its final blow in 1828, when *Karl von Baer* published his great work, *Developmental History of Animals,* which contained his careful observations on eggs and their developmental stages showing that differentiation proceeded progressively. Preformation soon faded away as a concept of the developmental process. Today, of course, the egg ins viewed neither as a completely undifferentiated, "*formless*" mass nor as containing a miniature, preformed individual. In the current view, sperm and egg contain a set of genetic instruction, genes, inherited from the parents. The genes may be thought of as providing the blueprints to be followed by the embryo in achieving its proper form.

NINETEENTH-CENTURY EMBRYOLOGY

With *von Baer's* discovery of the try ova of mammals in 1828 detailed study of the embryonic stages of development began, and *von Baer* can properly be regarded as the *"father of modern embryology."* Not only did he correctly identify the mammalian egg, but he described the *primary germ layers* of the embryo and the correspondence between the successive embryonic stages of higher organisms and the adult stages of more primitive organisms. Especially significant was his identification of the *notochord,* a basic embryonic structure characteristic of all animals (including ourselves) that belong to the phylum Chordata. It plays a fundamental role in establishing the organization of the early embryo. Perhaps von Baer's most significant contributions to embryology were his high standards of embryological investigation and his introduction of comparative studies on both higher and lower organisms, the essence of *comparative embryology.*

Drawing on von Baer's work, *Ernst Haeckel* (1834-1919) stated his biogenetic law, which may be epitomized by the phrase *"ontogeny recapitualates phylogeny."* This implies that a higher organism passes through a sequence of embryonic stages that reflect the adult stages of its more primitive ancestors. For example, clefts appear at an early stage in the pharyngeal (throat) region of mammalian embryos; these clefts bear a distinct resemblance to the gill slits of a adult fish. However, the pharyngeal clefts close up in the mammalian embryo, and the adult shows no trace of them. Likewise, the early mammalian heart is a tubular, two-chambered structure, like a frog's heart, before finally achieving the four-chambered condition seen in adult crocodiles, birds, and mammals. The notochord provides structural support in adult primitive chordates, such as the lancet *Amphioxus,* but is replaced by bone in the vertebrates. It persists as a significant embryonic structure in vertebrates, though only vestiges are found in the adult.

Catchy sayings like *"ontogeny recapitualates phylogeny"* easily become accepted as though they possessed deep meaning. However, they should not be regarded as revealed truths. A developing embryo does not truly *"recepitulate"* its evolutionary history; rather, the process of development requires that early stages be simple and later stages complex. A true understanding of the developmental sequences so well described by *von Baer* and his successors had to await the arrival of experimental embryologists, who go beyond simple description of developmental stages to a calculated interference with developmental processes.

THEORY OF PREFORMATION

In its crudest from the theory of *preformation* or *predelineation* postulated that the ovum contains in its substance a more or less perfect miniature of the adult animal, and that development consists of the growth and unfolding of this miniature creature into the adult form. In the case of man, this miniature organism was called *homunculus,* such a homunculus, which, by proper stimulation from the seminal fluid, would develop into the adult. At this point some of the preformationists got into difficulties in their speculations with regard to man, for, if there is a homunculus in the ovum, the homunculus itself must have an ovary whose eggs contain secondary homunculi, and these again must contain eggs that contain smaller, tertiary homunculi, and so on until the homunculi become so small that they necessarily had to vanish into nothing. To what absurdities some of the well meaning Naturphilosophen drove this argument is indicated by some of their speculations as to how many homunculi mother *Eve* carried in her ovaries. Their number was estimated to be two hundred millions, and it was believed by some that when these homunculi would be used up, no more human beings could be born because the supply would be exhausted and the world would come to an end. On this basis, even the year could be foretold when this event would occur.

Naive as these speculations may appear to us at the present time, they were put forth by men of fine intellects, capable of ken observation. An explanation of how these investigators could arrive at such bizzare conclusions may be found in their studies in the germination of seeds in plants, though it was not realized at that time that the seed was not at all homologous to the animal egg. Another factor leading to these conclusions was that most of these investigators used the chick egg, which was easily available and convenient to manipulate and incubate. Chick eggs are already well started in their early embryonic development when they are laid by the hen, and when they are then exposed to a temperature equivalent to that of the brooding hen they will continue their development. It is therefore not at all surprising that the great anatomist and zoologist *Marcello Maipighi* (1628-1694), who lived in Messina and Bologna, where in the summer the temperature may rise to over 100° F, did observe minute chick embryos in the unincubated eggs which he examined.

To this evidence on chicks were added the observations of *Jan Swammerdam* (1637-1680) and of *Charles Bonnet* (1720-1793) concerning insects. *Swammerdam* was careful observer who contributed a great number of facts to biology in his studies on insects. He, as well as *Harvey,* thought that the pupa (chrysalis) is the insect egg. *Swammerdam* could observe that in those form which undergo a complete metamorphosis in their development, such as flies and butterflies, the imago is ready formed beneath the skin of the young chrysalis. *bonnet* added to and strengthened this evidence by showing that the eggs of aphids develop inside the mother without fertilization, thus discovering the process of *parthenogenesis* and, at the same time, proved to his satisfaction that the ova of plant lice conform to the most exacting requirements of the preformation theory.

To the credit of *Bonnet* one should add here that he did not believe in the preformation theory in its cruder aspects. Though he did uphold the presence of a miniature adult form in the ovum, he thought that it would not necessarily have to be organized into a small replica of the adult. *Bonnet* was far too intelligent not to see the impossibility and the absurdity of such an assumption. He postulated that the preformation could be accomplished in the localization of centres around which organs and organ systems would form. This postulation comes quite close to our modern viewpoint of prelocalization of materials or organiforming areas in the ovum, to the so-called *mosaic ovum,* to be discussed later in the treatment of the mosaic theory. to the large number of brilliant defenders of the preformation theory the names ofthe great philosopher *Gottfried Wilhelm Leibnitz* (1666-1716) andthe renowned physiologist *Albrecut von Haller* (1708-1777) should be added.

An additional complication was added to the preformation theory when *Antony van Leeuwenhoek* (1632-1723) discovered the human spermatozoon to which he gave the name of animalcule, labouring under the misconceptiOon that it was an independent organism like a protozoon. In fact, he thought it was a parasite in the seminal fluid. when its significance in fertilization was finally established, the question arose as to whether the animalcule or the ovum contained the homunculus in man. This question was not as far out of place as one might asssume. Ever since, and even earlier than, the tiem of *Aristotle* (384-322 B. C.) it had been believed by many naturalists that the female contributed nothing, or very little, to the organization of the embryo and that the male alone was responsible for the generation ofthe offspring. Some considered the ovum, others the uterus as a mere accessory oran for development in that they contained the proper material environment in which the "*seen*" (seminal fluid) if the male found the proper place in which to develop, just as the seen of a plant has to find its appropriate environment inthe proper kind of soil in order to germinate. In fact, the word semen means seend and there are many allusions in the Bible and other ancient books to seed of men but never to the seed of women.

With this new aspect of the prefeormation theory, biologists studied the spermatozoa assiduously by means of the newly invented imcroscope until a microscopist by the name of *Hartsoeker* actually described human spermatozoa containing humunculi. Thus the preformationists became divided into two groups: the *animalcultists* and the *ovulists.*

Epigenetic Theory

Theories of preformation persisted well in the eighteenth century. By this time the German investigator, *Caspar Friendrich Wolff* (1759) offered an experimental evidence that no preformed embryo existed in the egg of cvhick. He suggsted that during embryonic development the ortgans formed successively in an epigenetic manner. He advocated that the future embryonic regions of an egg first consist of granules or the "*globules*" lacking in any arrangement that can be related directly to the form of structure of the future embryo. Later these "*globules*" were arranged into rudiments which can be called the *germinal layers. Wolff* saw that by the formation of local thickenings in some parts of these layers and thinning out in others by the formation of folds and pockets, the layers are transformed into various organs of the embryo. This method of progressive development from the simpler to the more complex, through the utilization of building units (granules or globules) is called *epigenesis.* However, the Wolff's theory of epigenesis was bluntly opposed by *Haller* and *Bonnet,* the two naturalists and philosophers of that time, but was confirmed by great embryologists of the early nineteenth century, notably *Von Baer* (1792-1876), the *father of modern embryology*. Later on after the discovery of protoplasm in animal cells by *Dujardin* (1835), this theory was spread out rapidly.

Today this theory is accepted, but not in the form offered by *Wolff.* The genetic constitution of an individual is fixed by heredity. On the basis of its genetic constitution, the individual will develop in a specific pattern. However, the external environment exetts its modifying influence upon that pattern. So long as the environment is normal for the developing ovum of a particular species, a normal individual will result. However, this normal is highly variable in its finer details because there are numerous delicate variations in the environment which escape an attention but they decidedly have modifying action in the expression of the genes.

Theory of Pangenesis

This theory was proposed by *Darwin.* It was mainly based on the cell theory pustulated by *Schileiden* and *Schwann* (1833). *Darwin* assumed that each cell ofthe body of an organism produces a minute copy of its own called *gemmule.* The gemmules are liberated into the blood stream. They are carried by the blood and deposited in the testis and ovaries. Thus the gonads are the *store-houses* of the various types of gemmules. The gemmules are then given to the gametes. The yuoung one formed from the gametes would be having all the gemmules of the parents. This theory supports the idea of inheritance of acquired characters (*Lamarckism-Effect of the Theory of Evolution on Embryology:* As a result of the recognition of these essential concepts, embryology became the common research field of the majority of biologists during the middle and latter part of the last century. However, the finest intellects of the biological sciences were not attracted to it until *Charles Darwin* (1809-1882) announced his theory of organic evolution in 1859. This golden age of embryological research brought to light untold treasures that acted as a stimulus to attract more men to the field, to add more to the store of knowledte in embryonic development, and to improve methods of attacking these interesting and absorbing problems. Thus the microscope was perfected to almost the limit of its resolving power; chemicals, such as salts, acids, and other agents, were developed for the delicate preservation and fixation of the embryos; and stains and dyes were discovered and evoved for the purpose of staining and differentiating tissues and organs. Haematoxylin, the most striking and useful dye in the whole domain of stains, was discovered in 1863 by *Waldeyer.* But the crowning effect in microtechnique was furnished by *His* (1831-1904) in the invention of the precision microtome in 180. His invention displaced the cumbersome and crude method of cutting thin sections with a razor, and allowed precise and accurate sections of hitherto undreamed thinness to be produced with ease. Whole embryos could now be sectioned into this slices, arranged in series, mounted on a glass slide, on which they were stained and then studied convniently under the microscope. It was truly a revolutionary period extending over a span of only a few decades. It is well to pause here fr a moment and reflect on those recent improvements and inventions which the average biology student of to-day takes for granted. In some instances they have been developed within the life span of persons still living.

All these factors, as well as many others too numerous to be discussed within the limits of this brief chapter, contributed to the great increase that was taken in embryological research and speculation. An outstanding figure of this period was *Ernst Haeckel* (1834-1919). Though his contirbutions to embryology were more philosophical and speculative than empirical, he ws a leading personality in biology. He was not only an indefatigable worker, but was able to stimulate the imagination of his students and colleagues and inspire them to greater efforts in their search for truth. His numerous students came top his from all over the world, and many of them have developed into outstanding biologists of international fame.

Recapitulation Theory

The recapitulation theory, usually ascribed to *Haeckel,* implies that the embryo of any species undergoes in its development the evolutionary or phylogenetic history of its race, or, stated more tersely, that "*ontogeny recapitulates phylogeny.*" Such a theory had already been advocated by *Fritz Muller* (1821-1897). He based it on the evidence discovered during his work on Crustacea, tracing the embryos of several species back to the Nauplius larva, which was common to all Crustacean species. This evidence was vigourously supported by *Haeckel* and was so much expanded by him that the recapitulation theory became associated chiefly with his name. The *biogenetic law* greatly stimulated embryologic investigation. It has been considered by many biologists, as for example by *Oscar Hertwig,* as *Haeckel's* most brilliant contribution to the subject. From the time of its announcement in 1868 to the beginning of the present century, it was the underlying theme whose recurrent intonation one could observe in the contributions to embryologic literature. As its most dominajt period its ever accumulating "*exceptions*" and defects were diligently overlooked, and its positive evidence was stressed with such great emphasis that a wholesome reaction was soon to appear in a new attack on the problem of development. The law of recapitulation, undoubtedly didmuch good in the latter quarter ofthe last century in the stimulation of thught, discussion and research.

Two of Haeckel's students who did a good deal of work of outstanding merit were the *Hertwig* brothers, *Oscar Hertwig* (1849-1922) and *Richard Hertwig* (1850-1937). In embryological research their more important contributions were their study of the origing of the mesoderm and the coelome, their attemts at artificial parthenogenesis, their first description of the fusion of the sperm and egg nucleus in fertilization, and their many other embryological and cytological researches. Others were struck by the divine spark of Hackel's genius and personality and became ardent defenders of his theories, speculations and ideas. Of tese men mention should be made of *Francis Maitland Balfour* (1851-1882), who in his short life span not only produced a prodigious amojnt of original work, but also organized the widely scattered accounts of comparative embryology into one science.

Beginning of Development Mechanics

Towards the end of the last century another trend made itself felt in embryological research—a trend which gave rise to two separate sciences, namely, cytology and experimental embryology. For the latter, the Germans coined the fortunate term "*Entwickelungsmechanik*" for which developmental mechanics is an approximate though inadequate translation.

This new turn began with a vigourous attack on the phylogentic conclusions of embryologists who saw in the development of the embryo a historical account of evolution; in other words, it was an attack on *Heackel's biogenetic law*. The man who led the attack, was *Wilhelm His* (1831-1904), the versatile inventor of the microtome. He pointed out that the causative agent for the development and differentiation of cells, tissues and organs could never be located by mere speculation along phylogenetic lines. He and several others were ofthe opinion that the formation and development of ortgans depend on what kind of functions these organs are called upon to perfom and not upon their phylogenetic history.

He advanced the bold idea that cleavage of ovum, blastulation, gastrulation and all succeeding stages in the development of the embryo could be explained on a purely mechanistic basis, and that any stage in developments is the immediate cause for the formation of the next stage, and that this stave, in turn, would act again as the cause for further and more advanced staves, etc. There is no beginning and no end to this chain, since it began with the activated ovum, fertilized or parthe-nogenetic, which had to come into existence by the union of an ovum and a sperm, or by parthenogenesis, and these germ-cells again had to come from gonads that were differentiated early in the parent organism.

The Germ Plasm Theory

This theory was proposed by *August Weismann* in 1904. It explains both the heredity and the development of an animal. According to *Weismann,* the body of an organism is formed of two types of cells, namely *somatic cels* and *germ cells.* The main bulk of the body is formed of somatic cells. The somatic cells disappear withthe death ofthe animal. So any change affecting these cells is not heritable. The reproductive cells (gametes) are said to be germ cells. The germ cells are carried tothe ddescendants, generation after generation. So any change affecting these cells in inherited. Each part of an animal is represented in the gametes by a separate particle called *determinant.* The determinants are located in the chromosonmes of nucleus just as the modern genes. The sum total of determinants represent the various parts ofthe adult organism. The complete set of determinants are handed down from generation to generation.

During cleavage the determinants are distributed unequally among the blastomeres. The distribution of the determinants is based on the future differentiation ofthe blastomeres. So each blastomere receives a particular type of determinant. Only the blastomeres which develolp into the gametes, receive the entire set of determinants since they are necessary for directing the development of the next generation.

Thus according to *Weismann,* each blastomere contains only one type of determinant. If a blastomere is separated from an early cleaving embryo, then, according to *Weismann,* the remaining cells should develop into only a particular type of organ and not into a complete embryo. But now it has been proved that a blastomere, separated from a four cell stage or even on eight cell stage of a sea urchin embryo develops into a complete larva. Thus this theory has no concrete evidence.

The Mosaic Theory

The views of *His,* especially his mechanical interpretation of development, were carried further and pushed with greater vigour by *Wilhelm Roux* (1850-1924) who had been one of Haeckel's students. He asserted boldly that we must look for the causes of growth and differentiation through the experimental method. On the basis of his work on the frog's egg, he came to the conclusion that complex differentiation can be reduced to simpler ones and that the latter can be, in the final analysis reduced to ordinary physico-chemical processes. Roux's outstanding contribution to embryology was his mode of attack in the solution of develompemtal problems. Just as his teacher, *Ernst Haeckel,* before him had stimulated his students bythe speculations of his fertile mind, so did *Roux* inspire his students and his followers by his unique and special methods of research. In his studies on the frog's ovum, he became convinced that certain areas of the egg are already destined in the ovary to develop into special regions. Thus, the dark areas of the animal hemisphere of the unfertilized frog's egg will develop chiefly into the head region of the animal, and the vegetal hemisphere into the posterior region. Prior to fertilization the entire ovum has a radial symmetry, while after fertilization has occurred, the meridional plane passing through the two poles and the sperm entrance point determines the right and the left side of the embryo and provides for it a bilateral symmetry. As cleavage continues and the blastomeres become progressively smaller, more areas for specific tissues and organs are delineated on the blastula so that, in the end, with regard to the potencies of its blastomeres, it comes to resemble a mosaic.

When the mosaic theory was tested on all types of eggs, it became evident that they did not all develop in the same manner. In the first cleavage of such eggs as that of Amphibia, *Amphioxus* or Echinodermata, the first two, and sometimes the first four blawstomeres, when separated, would give rise to two or (rarely) four complete larvae. Whatever the egg contained in the form of materials, it would give rise in the separated blastomeres to two complete larvae of this form. This formative stuff must therefore also be present in each of the first two blastomeres and also in the ovum. This was not the case inthe eggs of Annelida, Mollusca and most of the Arthropoda. When the first two blastomeres were separated in such eggs, they developed into monsters which could be idnetified in their later development as anterior or posterior portions of the normal larva. Eggs of the former group were said by *Conklin* to have "*indeterminate*" cleavage and the latter had "*determinate*" cleavage.

This was all very mystifying but the solution to the problem was very simple when theproper eggs were used for observation. Fortunately, eggs were found in either group which were marked into areas or zones by pigment, clear or corse protoplasm, yolk or other visible differentiations. The echinoderm and the amphibian egg cleaves in such manner that the first two or four blastomeres obtain equal amounts of any area, and the blastomeres are therefore supplied with every formed stuff and are enabled to form total larvae. In the eggs of Annelida and Molusca, on the other hand, the first cleavage does not distribute an equal amount of the matyerials of the egg into the first two or four blastomeres, and development, when blastomeres are separated, is therefore abortive.

Roux was forcefully opposed by *Oscar Hertwig, Hans Driesch* and others because he considered life on such a simple material and mechanistic basis, but many other investigators who studied the same processes in other eggs confirmed *Roux's* fundamental concept. This concept was the basis of *Roux's mosaic theory* of development, which, as we shall see later, contained a great deal of truth and furnished the support for present-day teories of development. Development, as upheld by *Roux* and as observed by him and othes, consisted of differentiation of cells from a more or less undifferentiated and homogeneous egg with crude and simple prelocalization of materials, to greater differentiation and specialization of cells as development went on.

Regulative Theory

Three years after *Roux's* experiment on the frog's egg, another German scientit, *Hans Driesch* (1891), performed a somewhat similar experiment on the sea-urchin eggs. At two-celled stage, he isolated the blastomeres and expected each one of them to develop into one-half blastula and ultimately one-half larva. But beyond his expectations, the one-half blastula, closed itself into whose blastula, whole gastrula and finally whole larva of one-half size. He repeated the experiment with 4, 8 and 16-cell stages and obtained the whole larvae by *Endres* (1895) and *Spemann* (1901, 1903) working with eggs of newts and by Schmidt (1933) on the frog's egg and obtained identical results.

Driesch reasoned that contrary to the theories of *Weismann* and *Roux,* the early cleavages of the egg are equatorial, "*a quantitative division of homogeneous material.*" Hence at an early stage, the blastomeres have equal potentialities. Their fate, he said, is determined by their position in the whole. Thus, he regarded a cleaving egg, as a harmoninous, equipotential system in which every part has potentially the properties of the whole. Development as observed by *Driesch* in sea-urchin eggs was to be called regulative development, and the eggs which were capable of performing such a development were called regulative eggs.

Gradient Theory

Since the time of *Driesch's* discovery that sea urchin blastomeres of the two-cell or four-cell stage can develop into a whole larva when isolated, the sea urchin egg has been chosen for immense number of investigations, partly because the of challenging nature of the early discoveries and partly becasue the eggs are readily obtained in large quantities, are easily artificially, and are easy to maintain through the process of gastrulation to the pluteus larva stage. Various operative and chemical procedures have been employed in attemtps to analyze the developmental processes leading to or involved in the formation of the blastula, the gastrula, andthe actively swimming larva.

Such an experimental approach of *Bovery* (1901), *C.M. Child* (1940), *Horstadius* (1955), *Runnstrom* (1967) and *L. Josefeson* (1959) on sea urchin eggs have contributed a most important theory, the gradient theory to us. The gradient concept was introduced in 1901 by *Boveri,* who considered that thevegetal region of the sea urchin egg exerted a dominant control over more animal regions, the influence decreasing towards the animal pole. However, he could not give any plausible reason for his concept and it was *C.M. Child* (1940) who stated that a single physiological gradient, called metabolic axial gradient controlled morphogenesis in sea urchin. Using the rate of reduction of vital dcyes, particularly Janus green as an indicator of the rate of oxidative metabolism, *C.M. Child* showed, that, there exist differences in rate of general metabolism in the ooplasm of animal and vegetal hemispheres of oocytes, and cleaving eggs of sea urchin and starfish.

Further, such differences in the rate of oxidative metabolism were found to be extended in the plane of animal-vegetal axis, so Child's concept is often named as a metabolic axial gradient. It has been generaly found that cytoplasms of animal pole of egg divides with high frequency and this frequency gradually becomes low towards the vegetal pole. This fact indicates that ooplasm of animal pole has high metabolic rate and the metabolic rate gradually decreases towards vegetal pole and thus, forming a metabolic axial gradient.

Child's metabolic axial gradient theory, has though solved many problems regarding cdleavage, gastrulation and morphogenesis, but, recently it has been modified by *Horstadius* (1955) and *Runnstrom* (1967). According to their theory, a doublt gradient system in assumed to control morphogenesis and differentiation in the early development of sea urchins. At the animal pole (the "*most animal*" region of the egg), the power to form structures characteristic of that region is most intense; at the vegetative pole (the "most vegetavie" region), the power to invaginate and form structures normally arising there (archenteron, mesenchyme) is at a maximum owing to the interception of the gradients, characteristic levels arise in the egg and at each level the state is determined by the relation between animalizing the vegetalizing agent.

Each gradient (*viz.*, animal gradient or vegetal gradient) is considered to extend to the opposite pole, so that, gradients overlap through the whole egg, and to be quantitatively different in nature, based on differences in metabolism one suggestion was that the animal gradient might be cortical and vegetal gradient might be in the interior. Recently, *Horstadius* and *JNosefsson* (1969) have showed that specific animalizing and vegetalizing substnaces namely one containing tryptophan and the other a nucleotide, respectively are present in matrue unfertilized eggs. The general indications that the animal factor promotes syntesis of ribonucleic acid and proteins, while the vegetative factor seems to inhibit the onset of these processes for a while, so that the time pattern of early protein synthesis will be different in the two regions.

Spemann's Theory of the Organizers

The term *Organizer* was first applied to a limited area in the dorsal lip of the blastopore. It was this area which could induce the development of the neural plate, notochord and mesoderm (chordamesoderm) with all its differentiations. If transplanted into any other region of the amphibian germ it would actually cause the development of another larva. This discovery made by *Spemann* was an achievement of the first magniturde, which was so far reching and so revolutionary that he was awarded the Nobel prize in 1935. This organizer is now better known as the *chorda-mesoderman field* which is the first o primary field to exist in the amphibian germ.

Secondarily it gives rise to secondary organizers or subordinated fields, such as nose, eye, ear, gill and others. Each field may cause differtiation by induction. Fiends are strongest at their center and their powers diminish with their distance away from the center. Sub-fields represent the final splitting up of the general chorda-mesoderm field into smaller, highly specialized area which show a high differentiation of specific organ-forming cells. The possibility of the invetigators interfering with the various fields is greatest during cleavage and it diminishes gradually past the neurula stage as the germ enters the yong larval stage.

Not only is the organizer of the dorsal lip of the blastopore the earliest retion to be differentiated into an area of determinate activity with regard to specific cellular function, it is also the most important and most comprhensive organizer. At that time other areas of the embryo, if transplanted, show that they are not fixed and determinate as yet.

For instanfe, a piece of ectoderm from the future tail region transplanted to the area where the optic vesicle is going to develop will give rise to the crystalline lens. However, this capacity is lost after gastrulation has been completed. More and more areas become organized and the whole process of differentiation becomes established and fixed. On the basis of the various concepts and theories, we must assume that the protoplasm of the egg is organized.

Even when such organization is not apparent, we must assume that a chemical organization exists. This organisation in certain eggs influences the location of the karyo-kinetic axis in the fist cleavages; it also influences the distribution of the ooplasm, or the ooplasmic inclusions, which in turn will have a bearing on metabolism, respiration and the other activities of the egg. In future research we have to turn to this organization to reveal to us the mechanism of development and differentiation of theovum.

Much interesting work is being doe by a large number of present-day investigators who follow along this line, which was pointed out first by *Jacques Loeb* and found to be so fruitful and productive of results, among whom are *E. Bataillon, A. Brachet, M. C. Child, Ives Delage, Ross Harrison, Otto Mangold, J. Needhm, Hans Spemann, W. Vogt* and a large number of more recent and younger experimental embryologists. It seems that in the immediate future the problems of developmental embryology will be solved through biochemistry and biophysics.

EXTRA-EMBRYONIC ORGANS IN PLACENTAL MAMMALS AND HUMANS

The extra-embryonic organs which develop during the embryogenesis process outside the embryo's body have multiple functions which provide growth and development of the embryo. The extra-embryonic organs are amnion, yolk sack, allantois, chorion and placenta.

Amnion is a temporary organ which provides the water media for the embryo's development. It evolutional development was promoted by the migration of vertebrates to the dry land. The inner surface of the amniotic vesicle is covered by the extra-embryonic (amniotic) ectoderm. Its external surface is covered by the extra-embryonic mesenchyma. They form the connective tissue component of amnion. By the end of the 7-th day of embryo's development the amnion's connective tissue contacts with the connective tissue of chorion. The epithelium of amnion grows over the amniotic stalk that is turns into the umbilical chord. At the early stages the amnion is covered by the simple squamous epithelium with an active rate of cleavage. It turns into the prismatic epithelium gradually. The microvilla are formed on the surface of the epithelial cells. The cytoplasm of these cells contains the small drops of lipids, glycogen granules and glycosaminoglycans.

The main function of amnion is the production of amniotic fluid that provides the medium suitable for embryo's development and protects t from mechanical damage. The epithelium of amnion reabsorbs the amniotic fluid. *Yolk sack* has appeared as an organ of nutrients storage (yolk). These nutrients were required to provide the development of embryo. Om humans the yolk sack is formed by the extra-embryonic endoderm and the extra-embryonic mesoderm (mesenchyme). In birds the yolk sack is filled with yolk. On the contrary, in humans and other mammals it contains serous fluid and, in fact, has a very small amount of yolk. It is formed at the 2-nd week and wouldn't participate in the process of embryo's nutrition, because at the 3-rd week the contact of the embryo with the maternal organism establishes and the hematotrophic nutrition begins. Blood insulae and primary blood vessels appear in the yolk sack's wall. They provide the conduction of oxygen and nutrients to the embryo. The yolk sack hemopoetic function is maintained up to 7-8 th week. Then the yolk sack undergoes regression and occupies the umbilical chord as the thin tube, which serves as the conductor of the blood vessels to the placenta.

Allantois is a small finger-like protrusion located in the caudal part of the intestinal endoderm and the visceral layer of mesoderm. It is the derivate of the yolk sack and consists of extra-embryonic endoderm and visceral mesoderm. The embryo is located between the amniotic and yolk vesicles. It is connected to chorion by the amniotic stalk, which allantois is growing into. The proximal part of allantois is located along the yolk stalk, while its distal part grows into the space between amnion and chorion. In humans the allantois is not significantly developed. However, it provides the nutrition and respiration of the embryo, because the blood vessels are growing towards the chorion along it. It is the organ of gas exchange and excretion. At the 20nd month of embryogenesis the allantois undergoes reduction and becomes a part of the umbilical chord together the reduced yolk sack.

Umbilical chord connects the embryo and the placenta. It is the elastic stalk covered by the amniotic membrane. The basis of the umbilical chord consists of the mucous connective tissue with fibres – it is also called the Warton's jelly. It is formed by the extra-embryonic mesoderm of the amniotic stem with two umbilical arteries and one umbilical vein along with the rudiments of yolk sack and allantois.

The mucous connective tissue provides the elasticity of the umbilical chord and protects the umbilical blood vessels from compression. Therefore, it provides the continuous flow of nutrients and oxygen to the embryo. It also protects the embryo by blocking the invasion of extra-vascular harmful agents from placenta into embryo.

New provisory organs provide the connection of embryo and the maternal organism in mammals and humans. They are chorion and placenta. Chorion is formed by trophoblast and the extra-embryonic mesoderm. By the middle of the 2-nd week the embryo is staying deep in the endometrium and its trophoblast forms the primary villi – the branched protrusions of symplasto-trophoblast and cyto-trophoblast. After the embryo implants into the uterus wall, they provide the connection with the maternal organism. Since this time trophoblast is called chorion. At the beginning of the 3-rd week the extra-embryonic mesenchyma enters the villi of the chorion. It grows over all surfaces, including the inner surface of trophoblast, or chorion. Therefore, the connective tissue stroma of villi is formed, and they become the secondary epithelio-mesenchymal villi of chorion.

By the end of the 3-rd week the mesenchymal cells located in the chorionic villi form the blood vessels. Their structure is similar to the one of blood capillaries. They contact with the blood vessels of chorion's mesodermal plate. The villi containing the well-developed blood vessels are celled tertiary. They form the structural and functional unit of placenta, that is called cotyledone. The chorion's villi are sunk into the lacunae in the endometrium. They are surrounded by the blood coming from the damaged blood vessels of the endometrium. Therefore, the utero-placentar circulation is formed and the hematotrophic type of embryo's nutrition is set. The chorion's villi grow bigger and become branched. The chorion's surface turned to the endometrium (maternal blood is coming from that side constantly) is covered by the branched villi that grow intensively and later form the so-called branched chorion. The surface of the chorion facing the uterus cavity is covered by villi that undergo gradual reduction and form the smooth chorion. The branched chorion and the neighbouring areas of endoometrium form the new extra-embryonic organ – the placenta, which provides nutrients and oxygen to the developing embryo. The terminals of the villi form the anchors that are fixed to the endometrium. Therefore, these areas of villi are not covered by the symplasto-trophoblast. They are covered by the cytotrophoblast only, that provides tight fixation.

The different kinds of animals have various types of placenta. The types of placenta structures are as follows:

1. *Epithelio-chorionic placenta* – the chorionic villi grow into the uterus glands and contacts to their epithelium (horses, pigs, dolphins, whales). That contact provides the chorion with proteins from maternal tissues and the fission of these proteins up to the level of amino-acids.
2. *Desmo-chorionic placenta* – its chorion partially destroys the uterus' glands epithelium and grows into the underlying connective tissue (cows, sheep). These types of placenta provide the development of the embryo up to the level, when independent feeding and mobility become possible right after birth.
3. *Endothelio-chorionic place*nta – the chorionic villi destroy the epithelium and the connective tissue of uterus. The contact the vascular endothelium (predators: feline, canine, walruses and seals).
4. *Hemo-chorinonic placenta* – the wall of uterus's blood vessels is destroyed, and the chorionic villi contact the maternal blood directly (humans, primates). These two types of placentas provide the embryo with maternal proteins, which are required for the development of tissues.

Functions of placents: trophic, respiratory, excretory, endocrine, protective, etc. Amino-acids, glucose, lipids, electrolytes, vitamins, hormones, oxygen, drugs and viruses cross the placenta and enter the embryo's circulation.. The embryo excretes carbon dioxide and metabolic waste into maternal circulation.

As a result of complicated development process all provisory organs get developed by the fifth week. They provide all the conditions required for the development of organism. The processes of organo- and histogenesis start.

NERVOUS SYSTEM AND SENSORY ORGANS EMBRYOGENESIS

Nervous tissue is being developed from ectoderm. The inducing activity of the notochorde, which is growing forward, promotes the formation of the thickened area in the medium part of th ectoderm. It grows forward from the primary node and is called the neural plate. At the 18-th day the neural plate starts to bend, so that the neural groove and neural cylinders are formed. Later, the 4-th week the sides of the neural groove move closer to each other gradually and join together, thus separating from the skin ectoderm. However, the junction of sides along the groove is not simultaneous – first it occurs at the neck area, then – at the caudal area and after all, the merging at the cranial part occurs. After all, the unpaired neural tube is formed. The formation of the neural tube is accompanied by the formation of neural folds, or the thickened areas that are located between the neural groove and the skin ectoderm. As the neural tube separates from the skin ectoderm, the cell clusters of the neural folds get isolated and form a solid layer between the ectoderm and the neural tube. It is called the neural plate or the neural crests. Some cells of the ganglious plate migrate into various directions thus forming the neural gangles. The material of the neural plate itself gets segmented and forms the spinal nodes. Therefore, the subdivision of nervous system into three parts occurs at the early stages of embryogenesis. These parts are head, medial and caudal regions. The fastest development rate is found at the medial region, where the spine is formed. Later, the head region develops into brain. The development of the caudal region into the neural tube occurs later, then the development of the upper areas. Its development is hampered early – actually it is the source of the caudal areas of the spine, which are characterized by low grae of differentiation.

DEVELOPMENT OF SPINAL CORD

The predecessor of the spinal cord is the tube which consists of cylinder cell. These cells undergo intensive mitotic cleavage. As a result, the walls of the neural tube thicken and become multilayered. At the transversal cut of the neural tube 4 layers can be found: dorsal (or the cover plate), ventral (or the bottom plate) and two lateral plates. The thick lateral sides of the neural tube are divided by the longitudinal stripe into the dorsal plate and ventral (or basal) plate.

Later the cells of the neural tube enter two lines of differentiation – some of them become spongioblasts, which are the source of neuroglia; the other become neuroblasts – later they develop into neurons.

At this stage the neural tube consists of three layers:

1. The inner, or ependymal layer, that covers the neural tube channel from inside
2. The medium mantle layer which contains neuroblasts and the neurons, that are developed from neuroblasts. It also contains the primitive neuroglial matrix, which is the progenitor area for neuroglial cells (astrocytes and oligodendrocytes)
3. External layer, or the marginal mantle. It contains no neuroblasts but is formed by the branches of the of the cells located in the ependymic and mantle layers. These branches are later developed into the neural pathways of the spinal cord, or the white substance. Later the cells of the ependymic layer turn into the cylinder-shaped ependymic cells, or glial cells that cover the channel of the spinal cord from inside. The mantle layer develops into the grey substance

with a groups of fast-growing cells. These are the progenitors of the motoric nuclei. Their neuritis gro from the spinal cord to the periphery thus forming the radices of the spinal nerves.

At the same time the walls of the neural tube grow irregularly. The cover plate and the bottom plate grow slower. Their width is insignificant until the very end of the development.

As the spinal cord develops, the regions of dorsal and ventral plates bend, while its right and left regions move closer to each other. Therefore, the glial and connective tissue septa is formed in its posterior part, while the narrow fissure develops in the anterior area. The neurites of the sensory cells, located in the spinal ganglia, enter the posterior horns of the spinal cord. These neurites form the posterior radices of the spinal cord.

FERTILIZATION DEVELOPMENT

The process of development begins with the fusion of gametes: egg (ovum) and sperm. The motile sperm swims to the egg, pierces its cell membrane and enters the cell. Fertilization is the fusion of the nuclei of the egg and sperm, and the single cell that results from this fusion is called the fertilized egg or zygote. During fertilization, the genetic material of the sperm and egg are combined.

Each gamete is haploid, that is, it contains one-half of the normal number of chromosomes for the species. At fertilization, the gametes combine to produce a zygote with the full number of chromosomes for that particular species. Another way to say this is that fertilization restores the diploid number. For example, haploid human gametes have twenty-three chromosomes; when the egg and sperm fuse, the diploid state of forty-six chromosomes is restored to the zygote. Thus, each parent contributes one-half of the chromosomal complement of the new individual, resulting in a new organism with genetic characteristics of both parents. This single cell, the fertilized egg, gives rise to all the organs of the individual—muscles, brain, liver, eyes—through highly regulated and timed processes.

The stages from fertilization to the birth or hatching of an individual organism are identical to those of all individuals of the same species. Developmental stages include rapid cell division or cleavage; formation of the germ layers through the process of gastrulation; and differentiation and growth of the organs and organ systems. These stages are categorized as the periods of embryogenesis (cleavage and gastrulation) and organogenesis (formation of organs and organ systems).

Cleavage

As soon as the sperm enters the egg, the cell membrane of the egg undergoes changes that prevent the entrance of additional sperm. Meanwhile, the chromosomes from each parent come together and, within a few hours, the first cell division begins. The egg degrades the cytoplasm and organelles of the sperm; only the chromosomes of the sperm contribute to the fertilized egg. The early cell divisions of the fertilized egg are called cleavage. The fertilized egg divides into two daughter cells called blastomeres. These two blastomeres divide into four blastomeres, the four blastomeres divide into eight, and so on.

During cleavage, the total number of cells increases, but the size of each cell decreases. The reason for this strange situation is that cell division occurs so rapidly that there is not enough time for the individual cells to grow bigger. The constant doubling of cells during cleavage results in a multicellular embryo very quickly. In a short period, the embryo has over one hundred cells arranged as a solid ball of blastomeres called a morula. The cells of the morula rearrange themselves into a single layer of cells surrounding a fluid-filled central cavity; the embryo at this stage is called a blastula.

Gastrulation

The next step in development is the formation of the gastrula by invagination, the folding in of the cells of the blastula at a point called the blastopore. The resulting gastrula is a double-layer cup of cells. The outer layer of cells is termed the ectoderm and the inner layer of cells is termed the endoderm. The inner endodermal layer surrounds a new cavity, the primitive gut. A third layer of cells, the mesoderm, develops between the ectoderm and endoderm in most animals. Ectoderm, mesoderm and endoderm are the three germ layers from which all cells, tissues and organs develop. Cells of the ectoderm differentiate into the epidermis, hair, nails, claws, sweat glands, tooth enamel, brain, and spinal cord. Mesoderm differentiates into muscles, blood, blood vessels, heart, spleen, reproductive organs, and kidneys. Endoderm differentiates into the cells lining the digestive and respiratory systems, the liver, gallbladder, and pancreas.

Induction

One of the more fascinating aspects of development is the determination of body form, pattern, and differentiation. Put simply, how does a cell know what it is supposed to grow up to be? How do cells of the endoderm know they are supposed to form the digestive and respiratory systems? Induction is the process during which individual cells are "told" what they are supposed to become. Hans Spemann (1869–1941) received the Nobel Prize in 1935 for over twenty years of research on development in amphibians.

In a series of elegant and delicate "baby hair loop" experiments, he demonstrated that when cells invaginate during gastrulation, they are induced to form specific cells and organs and that the primary inducer is a specific region of the blastopore.

Spemann tied a strand of baby's hair around a fertilized newt egg so that the nucleus and some cytoplasm were on one side of the ligature while the other side contained only cytoplasm. After several cell divisions, Spemann loosened the ligature and allowed a nucleus to pass over into the other side. When cell divisions began on the side with the transported nucleus, the ligature was again tightened to separate the two masses of cells. The result was the production of two newt larvae, one a bit older than the other.

These experiments demonstrated that all the nuclei of an early embryo are capable of producing embryos. This ability is the basis for much current practice of cloning agriculture or lab animals. In later experiments, Spemann found that the location of the ligature was important. If the ligature were placed so that each half of the fertilized egg contained the certain cells (called the gray crescent because of their colour) from the region destined to become the blastopore, two newts would develop.

However, if the ligature were placed so that the gray crescent was only on one half of the cell, that part would form a newt, but the half without the gray crescent would remain a formless mass of cells he called the belly piece. Further experimentation demonstrated that during gastrulation cells became committed to their developmental fates.

7

Agricultural Policies in Developing Countries

AGRICULTURAL POLICY

Since, it gained independence in 1947, India has pursued a policy of food self-sufficiency, primarily by boosting domestic production of its major food staples: rice and wheat. This basic goal has been achieved by developing and adopting high-yielding varieties, expanding irrigation, and increasing fertilizer use all aided by supportive output price and input subsidy policies. Cropped area has also increased steadily as expansion in irrigation has boosted multiple cropping and cropping intensity.

India's major policy instruments for domestic agriculture include minimum support prices (MSPs) for major crops; input subsidies for fertilizer, power, and irrigation water; and a Targeted Public Distribution System (TPDS) that provides subsidized food staples-primarily wheat and rice-to food-insecure segments of the population. MSPs are set annually and, if necessary, defended through market procurement by central and state government agencies. Historically, MSP policy for food staples has balanced producer incentives with the need to maintain consumer price stability, but recent adjustments to MSPs suggest a significant shift towards stronger grower incentives.

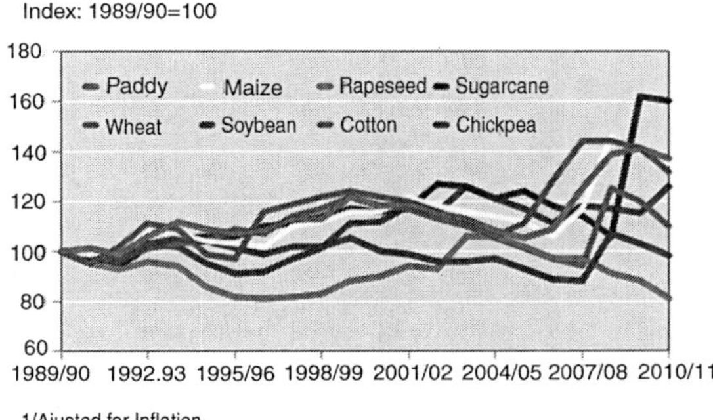

Fig. Real Minimum Support Prices for Major India Crops 1/.

Outlays for input subsidies for fertilizers, electrical power used for agriculture, and irrigation water distributed through surface irrigation systems increased about 11 percent annually in real terms (adjusted for inflation) between fiscal years 1993/94 (April-March) and 2008/09, when high world fertilizer prices pushed the total input subsidy bill to 1,609 billion rupees (US$35.04 billion).

On average, fertilizer subsidies, which include both subsidies to farmers and to the fertilizer industry, account for about 40 percent of all input subsidies. The cost of providing subsidized electricity for agriculture accounts, on average, for about 26 percent of total agricultural input subsidies, while subsidies to cover the operational costs of providing surface water irrigation typically account for about 21 percent of the total.

Billion 1999/00 rupees

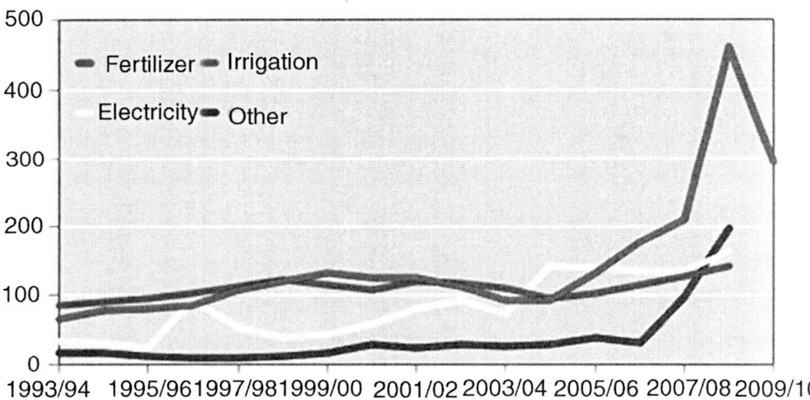

1/Value ajusted for Inflation.

Fig. India's Agricultural Input Subsidies 1/.

The budgetary costs of operating India's system of MSPs, public distribution, and storage for wheat and rice are accounted for in the "food grain subsidy." The real cost of the food grain subsidy increased sharply in the late 1990s, when India accumulated large surpluses of wheat and rice in government stocks.

The subsidy declined as stocks were reduced in the early 2000s but is now rising again due to hikes in MSPs, unchanged subsidized issue prices through the TPDS, and rising government stocks. Overall, the real cost of the food grain subsidy has increased about 9 percent annually since, the early 1990s, and is expected to reach a record of 513 billion rupees ($11.2 billion) in 2010/11. New food security legislation currently under consideration in India's parliament would significantly expand consumer eligibility for the subsidies under the TPDS in an effort to address the needs of India's large food insecure population.

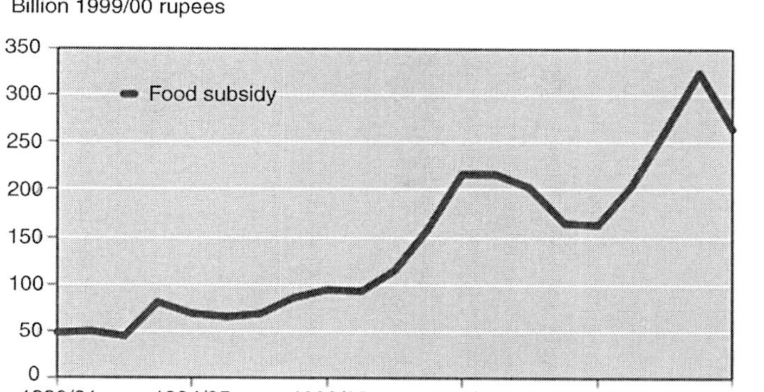

1/Values adjusted for inflation.

Fig. India's Food Subsidy 1/.

While India's policies have largely achieved the goal of self-sufficiency in staple cereals, the performance in other agricultural sectors has been mixed. Pulses (chickpeas, pigeon peas, lentils, dry peas, etc.) are an important protein source in Indian diets, but productivity is low and, despite having a liberal import policy since, the early 1980s, per capita supplies have generally declined. Recently, however, sharply higher MSPs have begun to boost output through diversion of land to pulse cultivation. In the oilseed sector, where India is among the world's largest importer of edible oils and an exporter of oil meals, India has traditionally provided high levels of protection for edible oils with the aim of boosting domestic oilseed production.

This approach imposed high costs on consumers in exchange for small gains in output. In 2007, vegetable oil tariffs were largely eliminated, leading to larger imports, lower consumer prices, and continued slow growth in domestic oilseed production. For cotton, the widespread adoption of hybrid Bt cotton - so far, the only genetically modified crop approved for cultivation in India - has led to rapid growth in output since, the early 2000s, supporting expansion of both raw cotton and textile product exports. India is a major global producer and trader of cane sugar, but both production and trade are highly cyclical, in part because domestic price policy has been ineffective in stabilising incentives to growers and processors.

With rising incomes diversifying food demand, non-staple foods -including fruits, vegetables, dairy products, eggs, and meats - are now experiencing the fastest growth in production and consumption, but with relatively little policy intervention in either input or output markets. India is the world's largest milk producer and output growth is being driven by strong consumer preferences for dairy products and, more recently, policies aimed at boosting private investment in dairy processing. Broiler meat and eggs are among fastest growing sectors of Indian agriculture, aided by consumer preferences for these sources of animal protein and the emergence of efficient private integrated poultry and egg operations that are both expanding supplies and reducing real prices.

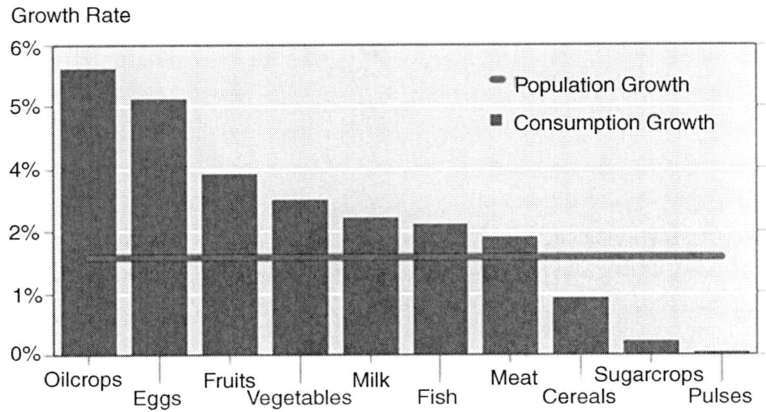

Fig. Consumption Growth by Food Group in India: 1991-2006 1/.

Stronger economic growth since, the major industrial and trade policy reforms of 1991-93 and slowed farm sector investment and growth continue to create pressure for change in India's agricultural policies. Food demand is expanding, diversifying, and placing new demands on underdeveloped agricultural markets and institutions. Longstanding production, procurement, and distribution programmes that focus on cereals are increasingly expensive and out of step with market demand. At the same time, rising subsidy outlays appear to be constraining public investment in transforming institutions and market infrastructure, while private investment in agriculture and agribusiness has been limited, in part due to national and state level regulatory interventions in agricultural markets. Although the central government plays an important role in recommending national agricultural policies and is the primary source of budget resources, agriculture is a "state subject" in the Indian constitution. As a result, all agricultural policies-including price supports, input subsidies, marketing regulations, and consumer subsidies-must ultimately be approved and implemented by state governments. The support of India's various states, which reflect a broad range of interests and resource endowments, is a key issue in reaching consensus on agricultural policy reform, and in implementing policy change.

There have been some important changes in agricultural policy since, the early 1990s, including the removal of quantitative trade restrictions completed in 2001, steps to reform state regulations that impeded private investment in agricultural markets, implementation of a unified value-added tax structure across Indian states, adoption of a new unified "Food Law" that will simplify food processing and trade, and large increases in the availability of credit for agricultural producers and agribusinesses. However, the poor performance of the rural sector was a key issue in the last rounds of national elections in 2004 and 2009, and remains an important and often contentious topic of public debate.

DOMESTIC AGRICULTURAL MARKET REGULATION AND INVESTMENT

Despite accelerating growth in the overall economy, rising consumer demand, and supportive price and subsidy policies, growth and investment in Indian agriculture and agribusiness have remained weak since, the early 1990s. There is evidence that numerous domestic policy interventions, along with weak infrastructure and limited institutional support for agricultural markets, have been a deterrent to new investment by agribusinesses and farmers. An array of central and state government policy interventions in domestic agricultural markets and industries have created disincentives and risks for investors in agribusiness, particularly larger scale and vertically integrated agribusiness enterprises.

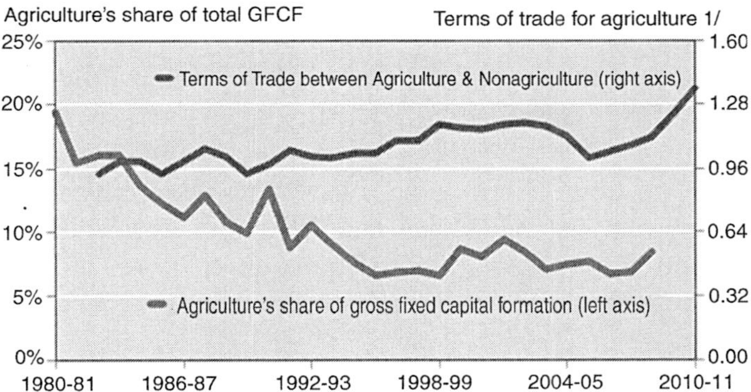

Fig. Agriculture's Terms of Trade and Share of Gross Fixed Capital Formation (GFCF) in India.

Interventions have included movement, storage, and private marketing restrictions for agricultural commodities, scale restrictions on agribusiness firms, high taxes on processed products, relatively high cost credit, and complex food laws. The climate for private investment is also shaped by weak transport and power infrastructure and lack of key market services, such as market information, risk management tools, and grading and inspection systems, which affect the private costs and returns of new investments. For farmers, disincentives have included trade and price policies that, despite MSPs, maintained relatively low domestic prices for many farm commodities, inefficient markets that dampen returns to growers, and the limited availability of public and private marketing services. Onfarm investment may also be constrained by India's large number of small and marginal farmers-accounting for about 39 percent of farm land-who often have limited access to input and output markets and more limited investment options.

Since, 2000, there is evidence that the policy environment is improving, and that investment in agriculture and agribusiness is beginning to show stronger growth. Aside from rising MSPs and input subsidies, a number of central and state policy changes appear to be stimulating more private investment. Recent changes include:

- Restrictions on private movement and storage on farm commodities are becoming less common.
- Restrictions that limited many types of agricultural processing to small-scale firms have been largely removed.
- State marketing laws are gradually being changed to permit development of private marketing channels.
- Taxes on processed agricultural products have been reduced and simplified.
- A new unified food law will simplify development of food processing industries.

Although power, transport, and other infrastructure problems will likely be solved only in the longer term, there is evidence that private investment is on the rise.

Recent private investment activity in food marketing ventures oriented towards development of supply chains and "front end" retail outlets represents a potentially important surge in private investment, as well as a turnaround in investor confidence in the future policy environment in agriculture and agribusiness.

AGRICULTURAL TRADE POLICY

From the time it gained independence in 1947 until the early 1990s, India maintained firm control over imports and exports of both agricultural and non-agricultural goods, making it one of the most closed economies in the world.

During this period, India's domestic markets were insulated from world markets by various restrictive practices, including trade bans, licensing regimes, state trading, and other quantitative restrictions, as well as high tariffs. Fundamental reform of domestic and trade policies affecting the industrial and service sectors began during 1991-93, when many restrictive licensing arrangements and quantitative restrictions were lifted and basic tariffs were substantially reduced.

While quantitative restrictions on imports of some agricultural commodities, most notably pulses and edible oils, were freed up in the 1980s and 1990s, the general removal of quantitative restrictions on agricultural trade occurred following the Uruguay Round Agreement on Agriculture (AoA).

The AoA required India to bind agricultural tariffs at ceiling rates ranging from 0 to 100 percent for primary products, 150 percent for processed products, and 300 percent for edible oils. In 1997, when India lost the balance-of-payments waiver that allowed it to maintain restrictive trade policies, India accelerated the process of lifting quantitative import restrictions (QRs). India completed the removal of QRs in April 2001, and nearly all agricultural products can now be imported subject to a tariff and to sanitary and phytosanitary standards.

While India's bound agricultural tariffs are among the highest in the world, applied tariffs for many products are set well below bound rates. India's policy has been to adjust applied tariffs periodically to help meet domestic price stability goals; either raising tariffs to help strengthen producer prices or reducing tariffs to help moderate rising consumer prices. In 2007 and 2008, India reduced tariffs for a number of major commodities-in a number of cases to zero-to help curb inflationary pressures and moderate the impact of high world prices on the domestic market. In general, however, tariffs have significant impacts on domestic prices of only the few agricultural commodities that India trades regularly in large volumes, such as edible oils and pulses.

For most agricultural commodities, Indian domestic prices are below transport-adjusted world prices. India's bound and applied agriculture tariffs.

India has also removed most quantitative controls on agricultural exports since, the late 1990s, but export controls are not restricted by the terms of the AoA, and India continues to impose bans and quotas on exports of agricultural commodities to meet domestic price policy goals. Recent examples include bans on exports of rice, wheat, and corn in order to ensure stable domestic prices when global process rose sharply in 2007 and 2008.

India also provides an array of policy incentives aimed at promoting exports of agricultural commodities, including the following:

- Import duty reductions or drawbacks for goods needed to produce processed products for export.
- Government support for domestic marketing and transport costs for exported products, consistent with World Trade Organisation rules.
- Government support for agroprocessing zones to produce goods for export.

AGRICULTURAL GROWTH AND PERFORMANCE

While the above analysis has provided a general view of the impact of economic reforms, this chaper examines agricultural growth and performance in the states of Bihar, Karnataka, Tamil Nadu and Punjab with their attendant policy implications.

The yields for various crops in these states differ greatly. While Tamil Nadu had the highest yield in rice, oil seeds and sugarcane, Punjab enjoyed the highest yields in wheat, coarse cereals, pulses and food grains. Karnataka on the other hand is seen to do well in cotton and Bihar performed quite well in pulses and coarse cereals. Further analysis and findings by Kalirajan and others (2001) show that Punjab had made remarkable achievements on the agricultural front while Bihar had remained stagnant in the last two decades, with Karnataka and Tamil Nadu showing moderate achievement. Clearly, differences in physical endowments, climatic conditions and institutional characteristics are some of the reasons for the varying productivity performance. Thus, having across the board economic reforms is likely to work less effectively than state-specific policy measures that enable each state's agricultural yields to reach their full potential. The comparative advantage of each state's agricultural production should be determined and with inter-state restrictions removed, total agricultural output would see a very significant increase.

For example, Karnataka with less favourable soil and water resources should be given incentives to concentrate on agro-processed products and corporate agriculture in horticulture, floriculture and animal husbandry, or to undertake watershed development to help with dry land agriculture. Many studies have indicated that with watershed areas, productivity growth has been mainly due to seed and fertilizer use. Thus, this state has to be given input subsidies for high yielding seed varieties but at the same time, the farmers need to be educated on the over use of chemical fertilizers.

With Bihar, agricultural performance is problematic on many fronts. First, although demographic pressure has increased and agricultural technology has improved, most of the uncultivated land is concentrated in southern Bihar, where irrigation facilities have not kept pace and the soil is of poor quality. Given the physiography of southern Bihar, wells are also unsuitable and thus the dominant mode of irrigation has been through tanks whose expansion and maintenance has been neglected. Second, the infrastructural facilities of Bihar have been lagging as seen by the infrastructure development. Due to infrastructural bottlenecks, availability of modern goods and services has not increased or their supply remains costly or unreliable.

Third, agriculture in Bihar is dominated by small and marginal farmers and the prevalence of mass poverty is largely related to the backwardness of agriculture. Fourth and importantly, the state agricultural policies in Bihar are in dire need of review. The semi-feudal production condition still exists in rural areas and the ineffective protection of tenancy rights has hindered agricultural growth. The slow pace of land consolidation reflects inadequate financial outlays and a shortage of manpower. Kalirajan and others (2001) note that marketing and extension services in Bihar are also rather weak compared to the other states.

Punjab on the other hand, was one of the few states which enjoyed the success of land reforms and the high priority of investment in rural infrastructure. Also, the irrigation base of the small and medium sized farms was comparable to that of large farms. In addition, the Punjab Agricultural University at Ludhiana contributed to the development of new seed varieties.

However, there are clear signs of a decline in crop yields since the 1990s and this has been associated with the increasing use of fertilizers and excessive water use which have increased the unit cost of production as a result of declining soil quality. Hence, care is needed when providing further input subsidies in fertilizer and water use. Another related fact is the steep increase in wages in Punjab and in the absence of productivity increases, the cost increase has affected the profitability of farmers. With Tamil Nadu, the main crop has been rice as this state is blessed with two monsoons. But from 1992-1997, there has been a steady decline in the areas irrigated by canals and an increase in well-irrigated areas while the use of tanks remains an unreliable source of irrigation. However, major improvements in about 10 rice varieties released in the early 1990s can be expected to improve productivity growth in rice production although pests and diseases as well as imbalance in the use of fertilizers are major constraints. Thus Tamil Nadu could do with subsidies of pesticides and farmers should be educated on the more effective use of fertilizers to obtain high yields. Interestingly, the cropping pattern of late has shown increasing substitution of food crops by commercial crops but there is concern that the benefits will reach farmers only with the development of adequate infrastructure such as roads and markets. However, that Tamil Nadu has a higher index than the all India average of infrastructure.

Challenges of Globalization

It is important to realize that globalization poses many challenges to a developing country like India, which had relied on a state directed and regulated policy regime for more than four decades. In moving to a more open, market-based economy there are many transitional problems that the country has to manage. The Government must play a pro-active role in facilitating the globalization process so that the opportunity sets for the economic agents are widened and the adverse effects of globalization are minimized. The Indian Government must also prepare the necessary information base and develop its capacity to articulate India's concerns and policy trade-offs in the international forums for multilateral trade and environmental negotiations.

In addition, the Government should embark on an extensive programme to educate farmers on the need to meet the standards required in the export markets. In fact, India needs to seek technical assistance in creating the capacity for meeting such standards and to consider watershed developments for environmental considerations. Equally important is the need to disseminate information about possible export markets to farmers, so that market access is achieved at minimum cost. Given the requisite information about markets and profitability, the likelihood of farmers investing in post-harvest and processing technologies and storage and efficient transportation arrangements as well as developing supporting infrastructure is very high.

Although the brave and bold move by India to reduce the tariff rate for agricultural products from 113 per cent in 1990-1991 to 26 per cent in 1997-1998 deserves to be applauded, the question of whether India is ready to compete in world markets remains to be seen. The infant industry argument may still hold for India to shield itself from external competition but one can easily question the length of time that is required to that end. Also, a delay in opening up to foreign trade has the danger that local producers may become too complacent and never be ready for competition.

As India opens up externally, it is also expected to face vulnerability in the wider international price fluctuations and thus Acharya (1998) claims that a minimum price support scheme is important. These prices can also act as a signal to adopt modern inputs and invest in yield-raising infrastructure for increasing production. For instance, keeping basic staple food grains at reasonable prices would induce farmers to switch over to high value crops. However, during the 1990s, Kalirajan and others (2001) shows that procurement prices especially for rice and wheat have been increasing faster than the general price level. Such high prices along with guaranteed purchases by the Food Corporation of India have pushed up market prices. These higher prices are partly responsible for the large buffer stocks with the Food Corporation. If this trend continues, India's comparative advantage will be eroded.

With openness and high price instability, unstable export revenue can also be expected. One way of reducing such risk is for India to diversify her agricultural exports. For example, since 1990, even in commodities such as tea, coffee, cocoa and spices, where India is supposed to have a comparative advantage international prices have been unstable. Besides increasing the type of exports to obtain more export revenue, India should also seriously consider exporting more value added agricultural products through agro-processing such as processed vegetables, fruits, fish and meat products given that export or even local demand for basic agricultural products would decline as incomes rise. The move to higher value added activities within the agricultural sector also spells greater opportunities for industrialization and vice versa as borne by Kalirajan and Shand's (1997) findings of a bi-directional relationship between agriculture and industry for most Indian states. On the other hand, Sivakumar and others (1999) establish empirical evidence of high forward linkages of agriculture due to the presence of agro-industries while Satyasai and Viswanathan (1999) show the significance of the spillover effects to the industrial sector via the intensive use of purchased inputs in the agricultural sector.

The lack or slow pace of internal or domestic liberalization is also seen to hinder the possible gains from external or trade liberalization. For example, although central zoning restrictions have been abolished, state government restrictions on inter-state and even inter-district restrictions on marketing and movement of goods still exist in many cases. This interferes with the benefits from crop specialization and economies of scale arising from comparative advantage. The land market is another example of distortion whereby land ceilings exist preventing the operation of large-sized farms. This has led to the emergence of a large number of small economically unviable land holdings. The easy leasing of land should be permitted with assurance of resumption. Yet another problem lies with the insufficiency of credit to agriculture. From 1995-1996, the Rural Infrastructure Development Fund was set up to allocate funds for the completion of projects and the government has committed itself to strengthening the cooperative credit structure through substantial refinancing and restructuring of the Regional Rural Banks. However, as mentioned earlier, due to varying institutional factors in the Indian states, these domestic reforms can be expected to yield quite different results.

Although India missed the opportunity to open up two decades ago, its attempts to do so now must be regarded as better late than never. Others such as Desai (1999) observe that, "the logic of the global economy as well as India's interests dictate that India become proactive in its liberalization policies. India must liberalize not because it has no choice but because it is the best choice".

His lament that India has adopted a 'victim mentality' when it really needs to adopt a 'winner mentality' has become less of a concern as over time, India has shown commitment to stay on the bandwagon of globalization. Having realized that globalization is a necessary but not a sufficient condition for high growth production, India has undertaken economic reforms, both internal and external. However, it must be ensured that these reforms are synchronized so that the pace of both reforms is set right in order to work hand in hand to promote agricultural productivity growth.

Thus, training the farmers and educating them appropriately to change their mindset and reorienting them to take up new activities or adopt foreign technology is of utmost importance. In this context, it is necessary to involve non-governmental organizations in training and mobilizing the rural poor to face the challenge of liberalization.

Also, with domestic economic reforms, more care needs to be exercised to draw up state-specific liberalization measures to maximize their benefits. Lastly, in the implementation of these reforms for successful globalization, one crucial element, not entirely within control is the need for good governance and stability in the political and economic environment. Political leaders who are the ultimate decision makers in these matters need to examine their own role dispassionately.

It is quite apparent that at this relatively early stage, there is little observable evidence of gains to India's agricultural performance after opening up. However, there could easily be benefits that have not yet surfaced, or are yet to be identified and perhaps too difficult or intangible to measure.

Whatever the case, it is highly likely that it is too soon to assess the full impact of globalization and economic reforms. Furthermore, the process of liberalization has been gradual and remains incomplete. For example, the complete removal of quantitative restrictions after March 2001 will have provided an opportunity for Indian farmers to tap world markets and, if they are successful, results should start to become evident soon. Export promotion via the development of export and trading houses as well as effective liberalizing export promotion zone schemes for agriculture are fairly recent measures and only time will tell as to how effective these measures are. Other possibilities such as agro-industry parks for promoting exports are also in the pipeline.

AGRICULTURE SECTOR

There were three distinct periods in the evolution of Indian agriculture and agricultural policy after Independence. The first period was before the Green Revolution up to the mid-1960s (Phase 1). The second period was the Green Revolution from the mid-1960s till the late 1980s (Phase 2). The third phase continues from then to the present day (Phase 3).

The first period saw the government focusing policy on the following:

- Irrigation,
- Land reforms,
- Community development, and
- Restructuring of rural credit institutions.

Land reforms involved abolition of intermediaries, tenancy reforms, acquisition of surplus land through ceilings on holdings and redistribution and consolidation of holdings; success varied across states. Land reforms quite successfully altered agrarian structures in states like Kerala, West Bengal and Karnataka, but failed to do so in states such as Bihar, Orissa and Rajasthan. Changes in agrarian structures occurred in Punjab, Haryana, Western Uttar Pradesh, Tamil Nadu and Andhra Pradesh.

A Community Development Programme was started in 1952, for integrated village development by coordinating the development of agriculture, animal husbandry, infrastructure and extension at the block level. Also in 1952 the National Extension Programme was spliced into the Community Development Programme, incorporating provision of technical inputs. The Intensive Agricultural District Programme (IADP), started in selected potential districts in 1960 and 1961, aimed to provide a package of high-yielding inputs (seed, fertilizers, plant protection measures, etc.) to farmers. The Intensive Agricultural Area Programme (IAAP), started in selected potential blocks in

1964 and 1965 aimed to provide technological inputs for identified crops. Whether these changes were successful or not depends upon how one benchmarks success. Compared to the low or negative growth in the period immediately preceding Independence, they were successful. Both productivity and output increased during most of the period. However, progress in productivity improvements was not commensurate with the requirements even during good monsoon years. There were institutional constraints in providing inputs like high-yielding varieties, fertilizers and irrigation, coupled with a switch during the Second Plan (1956–1961), with reduced investments in agriculture as fallout. In other words, the government was unable to inject the required financial resources, or develop adequate institutional back-up to enable widespread and sustained implementation. Moreover, the government was also unable to effect improvements in technology and the usage of improved seeds, inputs and methods did not occur until much later.

On the positive side the first phase did succeed in other aspects that facilitated later changes. Land reforms (although limited to some states and discussed elsewhere), assigning ownership rights to tenants, consolidation of holdings, among others, were important measures that started during this period and have continued to a varying extent to the present day. The establishment of many agricultural technology institutions and the birth of a widespread extension services system also transpired during this time (although it is now ineffective in many states). Most importantly, administrative systems were developed during this period with the capacity to take on the challenge of the Green Revolution when the technologies finally became available.

The mid-1960s witnessed a succession of exogenous negative shocks that had a major impact on agricultural output: a war with China in 1962, a war with Pakistan in 1965 and successive monsoon failures in 1965 to 1966 and 1966 to 1967. India had to resort to large-scale imports of food.

The Green Revolution package was based on increased availability/access to:

- High-yielding varieties (HYV) of seeds,
- Fertilizers,
- Irrigation,
- Biochemicals,
- Extension services,
- Availability of credit,
- Establishment of marketing and price-support mechanisms for farmers.

This was spliced with a price-support policy. The government's role in ensuring the availability of complementary inputs for HYV seeds was quite significant. Ahluwalia summarizes it well, " it required a comprehensive strategy for agricultural change requiring active Government intervention in many dimensions...".

A sustained effort was needed to expand irrigation with a shift from major to medium and minor irrigation projects. It was necessary to move the banking system into rural areas to provide credit for the purchase of biochemical inputs needed for HYVs. Nationalized banks were given the task of upgrading their rural operations and they succeeded to a large extent. Primary agriculture markets were regulated and some of the usurious practices of traders stopped. Extension services were set up and backed by numerous agricultural research establishments.

Development of appropriate varieties was critical given the heterogeneity in agroclimatic conditions. The ability of the government to effect adequate coordinating mechanisms was also quite important. Provision of credit was facilitated through the nationalization of banks. Fertilizer availability and accessibility was expanded through subsidies and was juxtaposed by public and private sector expansion in manufacturing. Moreover price support was instituted at remunerative prices.

This coordination role cannot be understated. While subsidized inputs were becoming available and productivity increases were rapid and concentrated, the price-support mechanism ensured that incomes for farmers increased. The three key characteristics of the mandate were: simultaneous handling of constraints in supply and marketing, a coordinated approach and recognition of heterogeneity in agroclimatic conditions.

This approach can be assigned most of the credit for the success of the Green Revolution. The growth rates in area, production and productivity of major crop groups. The asterisks in the table indicate the major contributing factors behind increase in production during that period. From 1949–1950 to 1964–1965, non-foodgrain was the major component of crop production and for both foodgrain and non-foodgrain, production growth primarily transpired through area expansion. This was reversed from 1967–1968 to 1980–1981. During this period, expansion of production for both foodgrain and non-foodgrain occurred through productivity increases, area expansion having slowed down. This trend continued from 1979–1980 to 1989–1990, with productivity increases becoming sharper and the area under foodgrain actually declining. However, in the last few years, productivity growth has slackened, and so has growth in production, particularly of foodgrain.

However the Green Revolution had several positive aspects: (1) As foodgrain production increased faster than the population growth rate, per capita availability of food increased. Per capita net availability of cereals and pulses was 394.9 grams in 1951, 468.7 grams in 1961, 468.8 grams in 1971, 454.8 grams in 1981, 510.1 grams in 1991 and 416.2 grams in 2001 (GOI 2004–2005). (2) Per capita income generation in agriculture increased. (3) Agriculture was better protected from the ravages of drought. (4) There was greater commercialization and diversification, with cropping patterns changing in favour of commercial crops and moving away from coarse cereals, even for small and marginal farmers. (5) Capital accumulation rose, including investments from the private sector. (6) While there may have been some reservations about the initial distributional impact of the Green Revolution package, the benefits became more broad-based in subsequent years.

Many holdings are marginal (less than 1 ha) or small (between 1 and 2 ha). Barring pulses and coarse cereals, all other crops seem to have benefited from the new technologies. In addition, there was a paradigm shift from foodgrain towards commercial crops and even fruits and vegetables, this diversification being at the expense of pulses and coarse cereals. Although a large percentage of small and marginal farmers is still dependent on foodgrain, there was some diversification among small and marginal farmers. For instance, the share of cereals in area cultivated by small and marginal farmers declined from 71.44 percent in 1970 to 1971 to 70.57 percent in 1980 to 1981 and 66.22 percent in 1990 to 1991; the share of fruit and vegetables increased in area from 2.43 percent in 1970 to 1971 to 3.25 percent in 1980 to 1981 and 3.71 percent in 1990 to 1991 (Despande *et al.* 2004). Thus it can be argued that there has been a degree of commercialization and diversification among small and marginal farmers but this could have been greater.

However there are some qualifiers. The first is that the coordinated approach did not lead to the spread of the benefits of the new technologies across India. Only a few areas benefited. It is difficult for a central government to introduce measures and mechanisms that are fine-tuned to the requirements of each state and district. , Success was based on the term "provision" — *inter alia* provision of credit, HYVs, subsidized fertilizers and extension services. These had a negative long-term impact on the development of market-backed services and commodities. Lastly, the new technologies were biased towards high usage of water for irrigation. As many parts of the country were not irrigated, power-operated motorized tubewells sprung up across the country. Crops such as rice in Punjab and sugar cane in Maharashtra have become quite important even though surface water is limited in some of these areas. The result has been a rapid fall in water levels.

In the 1980s, when the balance between demand and supply of cereals (foodgrain) was in sight, the overall goal of agricultural strategy/policy shifted from "maximization of production of foodgrain" to "evolution of a production pattern in line with the demand pattern". The implication of this shift is that the emphasis of the policy shifted from foodgrain to other agricultural commodities like oilseed, fruit and vegetables. The shift helped to increase the output of non-cereal food items.

Diversification has been occurring in India since Phase 1, however, it became more prominent in the 1980s (but waned in the 1990s). There is little agreement on how much of this was due to the government efforts and how much transpired because of changing economic conditions. It would be fair to argue for oilseed that the shift from coarse cereals to oilseed in rain-fed areas was aided by (i) a protective trade environment; (ii) favourable price policy; and (iii) the connecting of the Technology Mission on Oilseeds. At the same time diversification was not merely attributable to greater area under high-value commodities; the improvements in yield also would have been generated by government-aided supply and technology-related factors.

The improvements in both production and yield for the crop sector. There were significant improvements for livestock in the 1980s — the value of the livestock sector in total agricultural output value increased from about 18 percent in the triennium ending (TE) 1980 to 1981 to 23 percent in TE 1990 to 1991 (although it remained stagnant at that level by TE 1997 to 1998. Almost 70 percent of the livestock sector's production (in value terms) is accounted for by the milk and products subsector; the role of "Operation Flood" cannot be understated.

The value of the fisheries subsector swelled by about 50 percent during the 1990s, although as a share of the total it declined from 1.3 to 1.0 percent of the total agriculture sector value during the period. There were significant and coordinated government efforts to increase fish production. The central government's outlays towards the fishery sector as a share of the total agriculture sector outlays were more than doubled from 2 to 3 percent in the 1970s to about 5.5 percent in the 1980s and 1990s. Production and development programmes were instituted in both marine and inland areas. Farmers' development agencies were established both for fresh and brackish water areas. These programmes included technology upgrading components, encouragement and involvement of the private sector in activities such as seed, feed and other inputs and also creation of suitable infrastructure for storage, transport, marketing and credit. By 1998/1999 more than 50 seed hatcheries at the national level had been established, fishery industrial estates had been created and 30 minor fishing harbours and 130 fish-landing centres had evolved.

Diversification in favour of high-value commodities is considered by some to have only partly benefited through government efforts. Both econometric and GIS-based studies also suggest that the presence of infrastructure aided diversification. Rao *et al.* (2004) found that in their sample, the areas close to large urban centres (and as a result with better connectivity to urban centres) could diversify more to high-value commodities. They concluded that urbanization is a strong demand driver for high-value commodities. In an earlier econometric study, Joshi *et al.* came to a similar conclusion — that diversification into high-value commodities was demand driven, unlike the supply driven Green Revolution.

Between 1990–1991 and 1998–1999, the annual average increase in yields was only 1.79 percent and 1.31 percent for wheat and rice, respectively. Despite the successes of the Green Revolution, there continue to be concerns with agricultural performance during Phase 3, the present phase of globalization and diversification, initiated in the 1990s, and some macro issues emerge.

The first macro point is the one already alluded to indirectly. Compared to the 1980s, there was deceleration of agricultural growth in the 1990s. The cuts across all crops and both production and yields. The quality of the system for collecting agricultural statistics and the extent of deceleration can be debated, but the conclusion that growth in the crop sector decelerated in the 1990s is fairly obvious. In any case, deficiencies in the quality of statistics are constant factors and cannot explain the deceleration. Indeed, growth in non-food and non-crop output was also faster in the 1980s than in the 1990s and growth in animal husbandry, poultry, dairy, horticulture and fisheries also slowed down. This deceleration happened despite the reversal of historical discrimination against agriculture. In the 1990s, the terms of trade moved in favour of agriculture and this was independent of the terms of trade measures used. There were several reasons for this reversal of trend. There were higher prices for rice and wheat through support/procurement policies and prices of other crops also increased. There is a positive correlation between procurement prices, open market prices and higher prices in the PDS (public distribution system). Simultaneously, because protection granted to manufacturing declined, the relative prices of manufactured products also declined. *A priori*, one should therefore have expected a positive supply response and increased capital accumulation. However farm profitability declined in the 1990s. Not all farmers had access to higher support prices for rice, wheat or sugar cane, there was a deceleration in yields and real input prices also increased, thanks to prices for fertilizers, power and diesel that were closer to market-determined prices. Higher cereal prices also contributed to increases in wage costs.

Consequent to per capita income growth, NSS data show changes in consumption patterns, with a decline in consumption of cereals in both rural and urban India, despite cereals still being important in the food basket. There has been a shift from consumption of coarse cereals to rice and wheat and also a shift towards consumption of fruit and vegetables and even fish and eggs. Some of these changes. Consumption patterns are changing, even for the bottom 30 percent of the population and the shares of non-cereal food (fruit and vegetables) and non-food items have been increasing in consumption baskets. Increased cereal prices and lower rural incomes may have depressed demand for both cereal and non-cereal food, but there is no denying that preferences are also changing.

The Reason for Deceleration

Why did deceleration occur? More importantly, why did deceleration occur when better technologies were more readily available? In answering this question, it is important to distinguish between generic problems that continue to constrain agriculture and issues that became constraints in the reform decade of the 1990s. The former constitutes part of the agricultural cum rural reform agenda, the latter is more specific. First, one confronts a diminishing returns argument, with total factor productivity declining, as opposed to an argument based on increases in input costs or a slower increase in output prices. This argument is usually applied to Punjab, Haryana and the western parts of Uttar Pradesh where agriculture has become overcapitalized. In these traditional Green Revolution areas, there are also questions about unsustainable practices like excessive use of water and imbalanced use of fertilizers. Land has become degraded. These are traditional arguments associated with the Green Revolution and concern reduction in soil fertility, excessive use of fertilizers and imbalance of nutrient content in the soil, problems related to biomass availability, genetic erosion, waterlogging and salinization, depletion of groundwater tables, imbalances in nutrient availability because of changes in cropping patterns and contamination of waterbodies and soil by pesticides and fertilizers. Then there is the issue related to lowering of farm profitability in the 1990s. While these are important topics, they are not very convincing in explaining the 1990s deceleration, unless one plugs in a regional dimension. In the 1980s, availability of power, irrigation and infrastructure helped the spread of the Green Revolution to the Eastern Region, particularly for paddy rice. Owing to power shortages, among other factors, this osmosis was less evident in the 1990s.

Another explanation for the 1990s deceleration is reduced public investments, especially in irrigation; this is directly reflected in the reduced share of capital formation in agriculture in the overall GDP. Moreover, in the 1990s, the quality of public sector agricultural research, technology development and extension services deteriorated.

While public investments in agriculture have declined, especially investments in research, private investments have not been able to compensate for the decline in public sector investments. Thus the burden of fiscal reform has been borne by agricultural investments, especially at the state level, including expenditure on R&D. More specifically, expenditure on research stagnated, while that on extension declined. The state withdrew from spending on agriculture, without establishing alternative institutional mechanisms. State governments lack resources, a problem that was aggravated in the 1990s, and adversely affected the development of agricultural infrastructure.

This can be juxtaposed with the failure of risk mitigation instruments to develop and general regulatory failure. There were no coordinated policies or crop adjustment programmes to enable the move from rice and wheat towards pulses, oilseed or other crops. Diversification led to risk and uncertainty and in the absence of institutions, this was difficult to handle. Within public sector institutions, there was deterioration in management and monitoring norms.

Hence, there were instances of adulterated fertilizers and seeds that were substandard. Whether extension departments of state governments are the best agencies for quality checks and inspections of inputs like fertilizers, insecticides, pesticides, feed and seeds is debatable. However, the quality of these extension services deteriorated in the 1990s.

While only tangentially related to the slowdown of the 1990s, there is the question of the form public investments in agriculture should take. The Steering Group on Agriculture and Allied Sectors (SGAAS) does not question this, but the Approach Paper to the Tenth Five Year Plan (APTFYP) does, "The policy approach to agriculture, particularly in the 1990s, has been to secure increased production through subsidies in inputs such as power, water and fertilizer, and by increasing the minimum support price rather than through building new capital assets in irrigation, power and rural infrastructure. This strategy has run into serious difficulties." The statement further identifies that the deterioration in state finances and the financial non-viability of the State Electricity Boards have meant that crowding out has occurred in public agricultural investment/expenditures for:

- Roads,
- Irrigation,
- Expenditure on technological upgrading,
- Maintenance of canals and roads,
- Expansion of power supply to rural areas not covered,
- Improvement in quality of rural power.

Moreover the combination of dependence on subsurface water and free/subsidized power is not only environmentally harmful but has also led to:

- Excessive use of water that produced waterlogging in many areas,
- Falling water tables.

At the same time, there is no evidence of any improvement in income distribution. Last, it is also increasingly being noticed that despite the best intensions of the policy there is a significant imbalance between usage of N, P and K fertilizers. Moreover, financial constraints are increasing both for the state governments and their State Electricity Boards (SEBs).

These problems are particularly severe in the poorer states. The equity, efficiency and sustainability of this approach are questionable. The subsidies have grown in size and are now financially unsustainable. Power continues to be supplied free for farmers in many states and is otherwise highly subsidized given the high cost of production of SEBs. According to one estimate energy subsidies (of which power is one component) accounted for about 10 percent of all the subsidies in the late 1990s. This figure is likely to increase subsequently.

Moreover, the small and marginal farmers have been less able to access credit despite the existence of a large rural credit system. Poor land records are endemic across rural India. Also much of the agriculture sector is in the unorganized segment. But the financial system is largely in the organized sector and requires credible land ownership records, systematic record keeping and formal contracts. This has contributed to a situation where the bulk of the small and marginal farmers cannot access formal credit institutions. In the 1990s, availability of credit for large farmers may have risen, but declined for small farmers. Moreover, although 18 percent of priority sector lending is supposed to be for agriculture, agriculture's share in net bank credit has been more like 12 percent.

In summary, there is unarguable stagnation in Indian agriculture and more so in the foodgrain sector. This is due to a combination of factors ranging from overdependence on the government in some areas/sectors, falling ability of the various government entities to subsidize and/or directly provide various inputs and services and overuse of resources. The "mission mode" that was so successful in tackling the problems of low production cannot be a vehicle to take on this set of problems. This requires efficient functioning of the price mechanism, smooth transactions between the various stakeholders, the alignment of incentives at the microlevel and fine-tuning of technology to the needs of the farmer. There are arguments for and against the current system of minimum support price (MSP) and subsidized inputs.

FUTURE EXPECTATIONS

Because of deceleration in the 1990s, expanding population and rapid economic growth, food security concerns have been raised. Food security at the national level need not be equated with 100 percent self-sufficiency in foodgrain production. Adequate foreign exchange reserves should instead be the issue. The mindset is largely a reflection of the food import scare India confronted before the Green Revolution. Future demand projections are contingent on assumptions made about population, per capita income growth, the time frame and hunger removal. The time frame tends to be 2020. Population projections for 2020 vary with 1.315 billion being closer to the mark now.

Consider the basic trends first. With economic growth, population increases and changing age distributions, food requirements will be higher. How they match up with expected production given current trends, and what implications this has for international markets.

Population

India's population in 2001 was 1.02 billion according to the Census of India, making it the second most populous country in the world after China (with a population of 1.28 billion). It has been estimated that in the next 15 years, the population will increase by about 23 percent to 1.24 billion persons. But estimates vary, some being as high as 1.4 billion. High growth rates can be ascribed to death rates being lower than birth rates. Between 2001 and 2010, almost 150 million more people need to be provided with adequate nutrition and between 2010 and 2015, another 83 million people need to be covered. Most GOI estimates do not go beyond 2015, but trends till 2020 are not likely to be any different.

Incomes and Expenditures

According to India's national accounts, total GDP in 2003 to 2004 was about US$600 billion (per capita GDP of US$540). Of this, personal disposable income was in the range of US$512 billion and 24.6 percent of this income was directed into savings by the household sector. During the postliberalization decade, from 1993–1994 to 2003–2004 the average annualized growth rate of India's GDP has been around 6.2 percent and this has been accelerating in the early 2000s.

The poverty has declined significantly, and is expected to continue to fall. The poverty line is defined by the GOI as the cost of a package of commodities (about 80 percent food items and 20 percent other essentials such as clothing) that can provide about 2 400 calories and 2 100 calories to an average Indian citizen living in rural and urban areas, respectively. The HCR (head count ratio) in India had fallen to about 26 percent in 1999 to 2000 from 39 percent in 1987 to 1988. Our own estimates are that if the GDP growth is sustained, it will fall to about 14 percent by 2010 and then to 8 percent by 2015. These are not very different from other estimates. Two further points need to be mentioned in this connection. First, income distributions are typically log normal and as the thick part of the distribution passes through the poverty line, it is possible for sharp reductions in HCRs to occur. Second, when the Indian poverty line evolved in India in the early 1960s, the presumption was that health and education would be merit goods, if not public goods. In either event, they would be provided by the government and need not be ingredients in private consumption expenditure. Hence, these are not included in defining or computing the poverty line. The 1990s, however, witnessed a switch from public expenditure to private expenditure in both health and education. It is thus possible that the Indian poverty line might be redefined at some time in the future. Notwithstanding this possibility, the proposition about decline in HCRs, assuming an unchanged poverty line, remains valid.

Both rural and urban areas have a very similar poverty scenario. However, India has a long way to go before the whole population has the means to consume the minimum required calories per day. During 1999 to 2000, more than one-fourth of India's population was below the poverty line and this amounted to about 260 million people at that time. If the rapid rate decline in poverty after 1991 were to continue, not only the percentage, but also the actual number of people below the poverty line is expected to be lower. The net impact on the demand for greater nutrition is obvious.

The Changing Food Basket

As incomes, and as a consequence expenditures, increase, not only are expenditures on food and agricultural commodities as a whole likely to increase, but the consumption basket characteristics are also likely to change. In both rural and urban areas, dairy products, meat, fruit and vegetables are likely to have a greater share of the additional demand being generated.

This is the standard picture across different countries, and is also reflected in the differing elasticities observed by various studies both for India as well as other countries. The income elasticities for five different income groups (quintiles 1–5) in urban and rural areas of India.

The food requirement is dependent on the consumption behaviour, which in turn is a direct outcome of income. With development, incomes are expected to rise and would thus impact consumption behaviour. The income elasticities are a measure of the future demand of various food items.

Quintile one (Q1) denotes the poorest 20 percent of the population and Q5 denotes the richest 20 percent. Income elasticities across all commodities, as well as quintiles, are higher for the rural areas as compared to urban areas. The elasticities also fall as income rises, both within urban and rural quintiles.

Wheat shows an elasticity of 0.50 and 0.32 for rural and urban areas respectively, in Q1. However, the elasticity decreases to 0.47 and 0.04 for the same areas in the case of Q5. Similarly, the income elasticity of rice in rural areas drops from 0.72 to a negative 0.21 as we move up from Q1 to Q5. A similar pattern is observed in the urban areas as well. The elasticities of milk and milk products, along with eggs, show very high values of greater than 1 for the lower quintiles, implying that the proportionate change in consumption demand is greater than the proportionate change in income. However, these elasticities also follow the general decreasing trend as we ascend to higher income groups. Elasticities for chicken and other meat are lower than those of milk and eggs. For rural areas, income elasticities for meat are greater than 1 across all income groups (1.25 for Q5), while they are less than 1 for all income groups falling in urban areas. Meat eating is common across India, however a significant share of the population is vegetarian. Even among those who consume meat, at certain times of the year a largely vegetarian diet is followed. Consequently, the income elasticities of meat products may not be as high as in other countries.

In both rural and urban areas, the own price elasticity declines in absolute terms as we move from the very poor to the non-poor section of the population. For cereals in rural areas, elasticity declines from 0.7 to 0.1, while it changes from 0.5 to 0.1 for the urban areas. Milk, edible oils and meat show similar trends in both urban and rural areas, although demand for milk shows a higher sensitivity to price changes. Price elasticity of demand for sugar in rural areas increases marginally from 0.7 to 0.8, before declining to 0.6 for the non-poor.

In urban areas, however, the elasticity shows a gradual decline as we move to higher income groups.

The demand for pulses is extremely sensitive to price changes in rural areas, with the elasticity being as high as 2.4 for the very poor in rural areas, but falls to 1.2 for the non-poor. For the non-poor section in urban areas too, the price elasticity is 0.9.

CONSUMPTION OF AGRICULTURAL PRODUCTS

Others have estimated the growth in food consumption and most have similar insights. Milk and milk products will see the largest increase up to 2020. Fruit and vegetables, sugar and meat and fish consumption will also increase significantly. The reasons are obvious in light. Income increases matched by the high income responsiveness of these commodities will be the driving force. Among cereals, the highest percentage growth will be in wheat, followed by rice. Coarse grains are not likely to grow as much.

The demand for rice is projected to grow from 78.3 million tonnes/year in 2000 to 118.9 million tonnes/year in 2020, showing a compounded annual growth rate of 2.1 percent. The main factor driving this demand will be the increase in population. Wheat demand rises from 54.2 million tonnes in 2000 to 72.1 and 92.4 million tonnes in 2010 and 2020 respectively, with a growth rate of 2.7 percent. The responsible factors will be the rising population and positive income elasticity for wheat. The demand for other cereals will grow by 0.9 percent, primarily on account of change in preferences as incomes grow. On the whole, consumption demand for all cereals will show a growth of 76 million tonnes from 2000 to 2020 at the rate of 2.1 percent, which is marginally higher than the projected growth rate of the population.

Demand for food items such as milk products, meat and fruit is expected to grow at much higher levels of 3 to 5 percent in the next few years. This is a natural result of growth in incomes following economic development. The estimated demand for foodgrain in 2020 is 240.6 million tonnes/year, growing from 155.6 tonnes/year in 2000 at the rate of 2.2 percent *per annum.* The income effect will manifest itself to raise the demand for milk and milk products to 166 tonnes, edible oils to 11 tonnes, meat and fish to 11 tonnes and sugar and gur (jaggery) to 25 tonnes/year.

PRODUCTION OF AGRICULTURAL COMMODITIES

India has one of the largest land masses of any country of the world, and a high proportion of its land mass is arable. Although most of the land depends upon the monsoons for irrigation, irrigated areas have been increasing and comprise about 40 percent of the gross cropped area. However, yields in India are among the lowest in the world, and therefore there is significant scope for increases in the coming future.

A yield comparison across countries shows how India's crop production fares *vis àvis* international benchmarks. India's figures are contrasted with crop yields in the United States, China and an average for the world. While these can be compared for a number of different crops, the results are remarkably similar, with a few exceptions. For cereals, coarse grains and pulses, both the United States and China are well above the world average. India has extremely low yields, even by developing country standards. For oil crops and primary fibre crops as well, India has yield levels below the United States, China and aggregate world levels. Across different product segments, a similar picture emerges. While the static picture of India's crop production is not encouraging, the indicators for the future provide grounds for optimism. The potential improvements made possible by improvements in technology, organizational expertise and policy redirection have lifted the yield levels of many of India's crops. From 1961 to 2004, commodities with the largest increases in yield were wheat, maize and jute, and rice to a lesser extent. Pulses, oil crops and primary fibre crops have had muted increases in yield.

On an even more cautionary note, in the last ten years, improvements in yield have plateaued for Green Revolution heavyweights like wheat and cereals. From 2001 to 2004, several crops even experienced an absolute decline in their yields. Given these indications, prospects for the future of Indian agriculture production are mixed. While yield levels seem to have plateaued, they are still significantly below competitive levels for most crops. While countries at the peak of the agricultural technology curve must pour funds into R&D to invent new ways to increase production, countries on the other side of the curve stand to benefit from implementing those methods that have already been pioneered. Catching up is always easier and for this reason alone, India should be able to realize fast and significant gains in the near future.

PRICE COMPETITIVENESS

Currently, there are some estimates of price competitiveness, the results often being functions of whether an importable or an exportable hypothesis assumption is used and whether one uses the nominal protection coefficient (NPC), the effective protection coefficient (EPC), the effective subsidy coefficient (ESC) or the domestic resource cost (DRC). A widely referred to study by Bhalla (2004) revealed that:

- Most crops except oilseed, some coarse cereals and sugar are internationally competitive;
- More crops would be competitive if developed countries withdrew domestic support to agriculture;
- However, import competitiveness becomes reduced overtime because:
  - Successive price hikes have led to very high prices for many

agricultural commodities. For example major crops such as rice and wheat became non-competitive during recent years because of a fall in their international price combined with a major increase in their domestic price owing to increase in MSP.

- Productivity growth has decelerated, moreover production efficiency is being affected by lack of implementation of technological innovations.
- A major thrust is needed in infrastructure investment in general and investment in science and technology in particular for India to maintain and enhance its competitiveness.

PRODUCTION, CONSUMPTION AND SURPLUSES

The GTAP model has been used to predict India's trade with the rest of the world including India's own production and consumption. India's GDP is expected to grow at more than 7 percent for much of the period under consideration. Moreover, the growth rate is expected to increase for the coming decade. As population growth has been falling owing to falling birth rates for the past two decades, the growth in labour supply would decline over the next decade and a half. Irrigated area is expected to grow, although not too significantly, as further increases will require significant public investments that show no signs of accelerating. Private efforts are likely to be the main driving force behind the 1 percent annual increase in irrigated area. Skilled and highly educated labour is expected to rise much faster than unskilled labour; this reflects the latest improvements in the educational achievements of a large segment of the population. The capital stock would have to grow rapidly as well, and is expected to increase as labour supply growth tapers off to sustain the high growth rates of 8 percent.

These GDP forecasts need to be qualified. The GDP forecasts, with acceleration from 6.73 to 8 percent are actually a worst-case scenario, although one must mention that the BRIC (Brazil, Russia, India, China) report, generated by Goldman Sachs, involves real GDP growth of slightly less than 6 percent. A more likely scenario is GDP growth of 7.5 percent, accelerating to 8.5 percent and a third scenario would involve 7.5 percent accelerating to 9 percent. Long-term GDP forecasts are rare. In the short term, the Tenth Five Year Plan (2002 2007) talks about 8 percent real GDP growth during the Plan, while the National Common Minimum Programme (NCMP) of the UPA government considers 7 to 8 percent.

Although long-term projections are rare, when they are made, most experts expect real GDP growth of 7 to 8 percent for India in the period leading up to 2020. Notwithstanding reservations about the speed of economic reforms, consequent to debates about liberalization in a democratic policy, these estimates of 6, 7 or 8 percent are probably underestimates, even if one ignores the exchange rate aspect. There are different ways to argue this out and all of these trends reinforce one another. Looking forward to 2015, an average savings rate of 30 percent is certainly plausible given the current rate of 28.1 percent rising. Foreign capital inflows are also increasing. There is no reason why the average investment rate should not therefore be 32 percent. The present incremental capital/output ratio (ICOR) is around 4 (though estimates vary between 3.5 and 4.6). Whatever the figure, there is no reason why this should increase significantly in the near future. Indeed, with reforms, competition and resultant efficiency improvements, the ICOR should decline. But even with an ICOR of 4, there is growth of 8 percent.

It is obvious that these quantitative forecasts have not considered the possibilities of future environmental degradation, unforeseen changes in economic and trade policy and of course technological changes. The growth in rice is going to be in the order of 1 percent *per annum* till 2010, for both production and consumption. For wheat however, the rise in incomes is going to create a marginally higher growth in consumption than expected production increases. For both oilseed and processed foods, consumption increases will outstrip production increases. Given current conditions and trends, similarly, for forestry products, as well as other foods, consumption increases are going to be far higher than likely production increases.

Overall, the patterns are quite unambiguous. Although production levels are likely to increase significantly, they are not going to be able to match the increases in consumption, at current and expected overall economic growth. This does not imply a fall in export earnings, as overall expected price increases due to the opening of international markets, as well as reduction of subsidies will tend to have a positive impact on prices. This is regardless of the temporary impasse at WTO on agricultural liberalization, which hopefully, will be temporary. The net export earnings for rice, wheat, sugar and other grains.

India is likely to maintain the status of an exporting country for rice and wheat. However for oilseed, India is likely to be a significant importing country in another decade and a half. It will also remain as a marginal exporter of sugar and coarse grains. Despite being an exporter of jute and cotton, India will be a net importer of plant-based fibre. Moreover, it will also start to become a net importer of other horticultural crops. Although overall milk production is expected to increase rapidly, consumption increases will prevent India from becoming a large dairy product exporter. Cattle and red meat exports are likely to increase, although other animal products, such as leather, are likely to see a net increase in imports.

EXPECTATIONS FOR INDIA

Overall, India will remain a marginal exporter of wheat, coarse grains, sugar, cattle and red meat, fish and other foods. However, it will become a significant importer of oilseed, forestry products, other animal products and plant fibre. For products such as plant fibre and animal products, its position as a net exporter of manufactured items will be facilitated by larger imports of raw materials. For the other segments, rising domestic consumption not matched by domestic supply increases, will be the driving principle.

Forestry products are likely to be a significant import item, the bulk being related to wood. Industrial and furniture requirements are the important components that will drive forestry product shortfalls. Although India has large forest cover, commercial forestry is insignificant and is unlikely to expand in a big way given its current environmental protection laws. The statistics for forestry products have large gaps (much more than is usual for India). According to some FAO estimates, production in India for industrial roundwood has been falling (it was about 1.6 billion m^3 in 2000), while fuelwood has stagnated at about 2.9 million m^3 . Pulp and matchwood have also been showing a negative trend. On the whole, it is expected that India will be substituting imports for domestic production more and more in the near future.

The overall position is quite unambiguous. For Indian policy-makers, the most worrying aspects are stagnating yields. This in itself is not surprising, given the low investments in agriculture, as well as the constraints on agricultural trade. If India is to become a significant exporter of other agricultural products apart from rice, emphasis will have to be placed on improving yields. Investments in rural areas are therefore essential.

How would these estimates change if growth assumptions were different? India would continue to be a large importer of forestry products whether economic growth is 6 or 8 percent. Although the quantum may differ. Similarly, its position as a significant textile and garment producer may require it to import plant-based fibre, in spite of plus or minus 1 percent variation in growth. (What the numbers in this case are more sensitive too is the productivity assumption. Yields are currently quite low in large parts of India and the introduction of new varieties, BT cotton being one example, may lead to a rapid increase in cotton production in coming years.) India's low production of oilseed and high requirements will also drive its oilseed imports. But this needs to be qualified. Given its land area, it is conceivable that productivity enhancements could lead to a lower shortfall in oilseed and pulses. However, current trends indicate this to be a remote possibility. Now rice remains. Compared to the past, rice is not a preferred cereal; income elasticity measures also indicate that at the middle and upper income levels, the parameters are lower. In other words, estimates of rice surpluses are not as sensitive to economic growth assumptions.

How would these surpluses affect India's trade partners? This requires an examination of which countries could potentially be India's major trade partners in agricultural commodities.

LINKAGES BETWEEN MACRO-ECONOMIC AND AGRICULTURAL PRICE POLICIES

The objectives-constraints-policies framework applies to macroe-conomic policy as well as to price policy. Common macro-economic objectives include rapid economic growth, a desirable distribution of national income, reasonably low unemployment, and moderate or low inflation. In addition to facing the sectoral constraints of supply, demand, and world prices, macro-economic planners are also confronted by a need to maintain an approximate balance in the national fiscal accounts (government revenues and expenditures) and in the foreign-exchange accounts (export earnings and foreign capital inflows versus import expenditures and foreign capital outflows). The macro-economic policies available to further these objectives in light of such constraints include fiscal and monetary policies, budgetary policies, and macro price policies influencing the foreign-exchange rate, interest rate, wage rate, and land rental rate.

The direct effects of macroeconomic policy on agricultural systems are felt through the macro price policies, especially exchange-rate policy. Fiscal and monetary policies influence agricultural systems indirectly by the interest and exchange rates. Budgetary policy-decisions on allocating both the recurrent and the capital budgets of the national government-also have indirect effects on systems, because budgetary choices influence agricultural price policy (through the availability of recurrent funds for subsidies) and public investment policy for agriculture. The three kinds of macro price policies affecting factor prices can be important in individual factor markets, although little can be said about them in general.

Some useful general lessons can be drawn from the relationships among fiscal and monetary policy, inflation, and the exchange rate and those between the exchange rate and price policies. Inflation is caused principally by macroeconomic policy-decisions to run fiscal deficits financed by expansionary monetary policy-abetted by inflation abroad that causes the prices of imports and exports to rise. If the government chooses to have a fixed-exchange-rate regime, the exchange rate will be changed only through discrete policy decisions, not because of market forces. When governments create inflation and then choose not to depreciate the nominal value of their currencies (by changing the exchange rate so that more units of domestic currency are required for each unit of foreign currency), profits are squeezed in agricultural systems that produce tradable commodities.

The real exchange rate becomes overvalued when the rate of depreciation is less than the rate of inflation. Overvaluation of the real exchange rate imposes an implicit tax on producers of tradables (by keeping the domestic currency prices of their outputs artificially low), forces farmers growing tradable food crops to pay implicit food subsidies that benefit consumers, and permits artificially cheap imported inputs. A policy creating inflation with fixed nominal exchange rates squeezes agricultural profits, transfers the burden of subsidizing food from the government treasury to farmers, and makes projects based on tradable inputs appear to be more profitable than they would be if the exchange rate were set appropriately.

This state of affairs can be corrected if a government chooses to change the exchange rate. Devaluations are often difficult actions to take politically, because their short-run effects usually benefit rural inhabitants who have limited political power and harm powerful urban interest groups. Some form of foreign-exchange rationing is inevitable when the real exchange rate is overvalued, and this rationing is most often achieved by quantitive restrictions on imports that compete with domestically produced manufactures. Politically powerful urban manufacturers and their employees then shift from being supporters of devaluation to being vocal opponents of it. The prices of their products are protected from the taxing effects of overvaluation by the import quota, and the overvalued exchange rate permits them to obtain tradable inputs at artificially low prices.

PRICE POLICY GRAPHS

A set of PAMs for the country's principal representative agricultural systems provides analysts and polcy-makers with informative pictures of the existing structure of policies affecting agriculture and with a useful analytic tool for investigating the effects of future policy change. However, in most countries, there is no information base to permit construction of historical PAMs that would show changes every two or three years as trends in world or factor prices and technologies changed.

Budget data might be available at best for a few systems during scattered years. But informed policy analysis requires an understanding of the recent history of policy changes as well as the detailed array of profitabilities in a given base year. This need can be met at least partially by the construction of price policy graphs.

A price policy graph is a device to permit easy visual comparisons of year-to-year movements in three price series-world prices (cif import or fob export, adjusted to a domestic wholesale market level), domestic market prices (at both the wholesale and farm levels), and domestic policy prices (guaranteed floor prices to producers and announced ceiling prices to consumers). Price policy graphs, based on annual data for fifteen to twenty years in the recent past, can be constructed for the principal agricultural commodities produced and for the main tradable inputs into agriculture.

They allow a quick visual review of the pattern of price levels and price stability. If historic price policy graphs are continuously updated, they can serve as particularly useful complements to PAMs in the presentation of policy analysis.

CONCLUDING COMMENTS

Several practical lessons for practitioners emerge from this study of agricultural policy analysis. Approaches to issues and the policy agenda can be organized within the objectives-constraints-policies framework, and diagrammatic analysis can be used to identify the general direction of policy effects.

Historical perspective can be provided through a compilation of price policy graphs for the most important agricultural products and inputs. Much insight is gained from using the PAM approach to the quantitive analysis of agricultural systems. The construction of PAMs, complemented by historical price graphs, provides essential baseline information for the analysis of agricultural policy.

The standard approach to agricultural policy analysis relies on estimated elasticities of supply and demand. When policies raise or lower market prices, use of the elasticities permits the analyst to quantify changes in amounts produced and consumed; income transfers among producers, consumers, and the government treasury; and efficiency losses or gains. The PAM calculations usually are based on budget data, not elasticities. A strength of the PAM method is the disaggregation of supply in terms of technology and agroclimatic zone. Such disaggregation permits a detailed understanding of constraints on systems and provides a basis for the analysis of investment and technological change influencing the dynamic comparative advantage of agricultural systems. The principal weakness of the PAM approach is that empirical applications may not correctly specify all the marginal adjustments to alterations in output and input prices. Without sufficient information (such as elasticities of output supply and input demand), exact PAMs cannot be constructed, and approximations must be made. Unless this is done, the empirical researcher will be left with nothing more than a numberless diagram, little understanding of how the many divergences affecting agricultural systems offset one another, and no input into the policy-making process. Budget-based PAMs fill this gap in agricultural policy analysis.

POLICY ANALYSIS MATRIX IN AGRICULTURE

This section explains the construction of the policy analysis matrix and the derivation of measures of efficiency and policy transfer used in agricultural policy analysis. The study of agricultural policy spans three levels-microeconomic behaviour of producers, marketing and trade, and macroeconomic linkages. Practitioners of agricultural economics typically give different emphasis to these three topics; micro production issues receive the greatest attention, marketing and trade get less, and macroeconomic links receive little or no coverage. This book argues that excessive specialisation precludes successful policy analysis; applied agricultural economists need to understand all of the components of and links among farming systems, domestic and international markets, and macroeconomic policy. Policy analysts have to appreciate feedbacks and tradeoffs within the big picture.

The PAM approach is a system of double-entry book - keeping. Analysts using PAM have to provide complete and consistent coverage to all policy influences on returns and costs of agricultural production. With this method, applied economists need to be equally capable of analysing, for example, fertilizer response functions, quantitative restrictions on trade, and real effective exchange rates. The main empirical task is to construct accounting matrices of revenues, costs, and profits. A PAM is constructed for the study of each selected agricultural system-using data on farming, farm-to-processor marketing, processing, and processor-to-wholesaler marketing. The impact of commodity and macroeconomic policies can then be gauged by comparison with the absence of policy.

PRACTICAL ISSUES ADDRESSED

Three principal issues - the impact of policy on competitiveness and farm-level profits, the influence of investment policy on economic efficiency and comparative advantage, and the effects of agricultural research policy on changing technologies - can be investigated with the PAM approach. The results can be used to identify what kinds of farmers - categorized by the commodities they grow, the technologies they use, and the agroclimatic zones in which their farms are located are competitive under current policies affecting crop and input prices and how their profits change as the policies are altered. This issue of farm policy-how agricultural prices affect farming profits - is of primary importance to ministries of agriculture. In the PAM approach, farm budget data (sales revenues and input costs) are collected for the principa, agricultural systems.

The determination of profit actually received by farmers is a straightforward and important initial result of the analysis. It shows which farmers are currently competitive and how their profits might change if price policies were changed.

A second issue concerns the economic efficiency (or comparative advantage) of agricultural systems and how additional public investment might change the current pattern of efficiency. In what commodity production systems, defined by technology and agroclimatic zone, does the country currently exhibit strong or weak comparative advantage, and how might new investments, using government revenues or foreign aid funds, improve this picture? Investment policy is of primary interest to economic planners who allocate capital budgets, including foreign aid, in attempts to increase efficiency and speed the growth of national income.

With the PAM method, the analyst reassesses the revenues, costs, and profits indicated in farm-level and marketing budgets. Efficiency valuations of outputs and inputs are meant to lead to the highest possible levels of national income. The difference between revenues and costs for a system-both valued in social prices-is social profits, a measure of economic efficiency. New investments that reduce social costs also increase social profits and improve efficiency. An understanding of the array of social profitabilities of agricultural systems greatly reduces the number of detailed benefit-cost analyses needed to evaluate investment alternatives.

A third and closely related set of issues is how best to allocate funds for agricultural research. How can economic analysis be used to help determine the most fruitful directions for primary and applied research to raise crop yields and reduce social costs, thereby increasing social profits? This question is faced by decision-makers in the international agricultural research centers, in several international organisations, and in the agricultural research establishments of certain countries. It is a question also asked by central planners who make allocations to agricultural research budgets.

The approach used in PAM analysis begins with the calculation of existing levels of private (actual market) and social (efficiency) revenues, costs, and profits. This calculation reveals the extent to which actual profits are generated by policy transfers rather than by underlying economic efficiency. Next, agricultural scientists need to project changes in yields and inputs resulting from alternative research programmes. The effectiveness of such changes can then be gauged by an examination of how they alter private and social profits of current technologies.

THE POLICY ANALYSIS MATRIX

The policy analysis matrix is a product of two accounting identities, one defining profitability as the difference between revenues and costs and the other measuring the effects of divergences (distorting policies and market failures) as the difference between observed parameters and parameters that would exist if the divergences were removed. By filling in the elements of the PAM for an agricultural system, an analyst can measure both the extent of transfers occasioned by the set of policies acting on the system and the inherent economic efficiency of the system.

Profits are defined as the difference between total (or per unit) sales revenues and costs of production. This definition generates the first identity of the accounting matrix. In the PAM, profitability is measured horizontally, across the columns of the matrix.

Profits, shown in the right-hand column, are found by the subtraction of costs, given in the two middle columns, from revenues, indicated in the left-hand column. Each of the column entries is thus a component of the profits identity-revenues less costs equals profits.

Each PAM contains two cost columns, one for tradable inputs and the other for domestic factors. Intermediate inputs-including fertilizer, pesticides, purchased seeds, compound feeds, electricity, transportation, and fuel-are divided into their tradable-input and domestic factor components. This process of disaggregation of intermediate goods or services separates intermediate costs into four categories-tradable inputs, domestic factors, transfers (taxes or subsidies that are set aside.

Table. Policy Analysis Matrix

	Revenues	Costs		Profit
		Tradable Inputs	Domestic Factors	
Private Prices	A	B	C	D
Social Prices	E	F	G	H
Divergences	I	J	K	L

Notes: *Private profits,* D, equal A minus B minus C. *Social profits,* H, equal E minus F minus G. 'Output transfers, 1, equal A minus E. 1nput transfers, J, equal B minus F. *Factor transfers,* K, equal C minus G. Net transfers, L, equal D minus H; they also equal I minus J minus K.

Ratio Indicators for Comparison of Unlike Outputs:

Private cost ratio (PCR): C/(A - B). Domestic resource cost ratio (DRC): G/(E - F) Nominal protection coefficient (NPC) on tradable outputs (NPCO): A/E on tradable inputs (NPCI): B/F Effective protection coefficient (EPC): (A - B)/(E - F) Profitability coefficient (PC): (A - B - C)/(E - F - G) or D/H Subsidy ratio to producers (SRP): L/ E or (D - H)/E in social evaluations), and non-tradable inputs (which themselves have to be further disaggregated so that ultimately all component costs are classified as tradable inputs, domestic factors, or transfers).

An example illustrates the process of disaggregating intermediate goods or services. Fertilizer is for most countries a tradable intermediate input. If a particular country is a net importer of fertilizer, the social valuation of a specific kind of fertilizer for its agricultural system is given by the cif (costs, insurance, freight) import price for that fertilizer plus the social costs of moving the input to the representative location in the system. Finding the import price is usually straightforward. Finding the social valuation of the domestic marketing costs is another story, however. It is necessary to study the transportation industry-road or rail-and disaggregate the costs into labour, capital, fuel, and so forth. Each type of cost then needs to be further broken down through use of an appropriate world price and an estimate of local transportation costs.

PRIVATE PROFITABILITY

The term 'private' refers to observed revenues and costs reflecting actual market prices received or paid by farmers, merchants, or processors in the agricultural system. The private, or actual, market prices thus incorporate the underlying economic costs and valuations plus the effects of all policies and market failures. The private profits, D, are the difference between revenues (A) and costs (B + C); and all four entries in the top row are measured in observed prices. The calculation begins with the construction of separate budgets for farming, marketing, and processing. The components of these budgets are usually entered in PAM as local currency per physical unit, although the analysis can also be carried out using a foreign currency per unit. The private profitability calculations show the competitiveness of the agricultural system, given current technologies, output values, input costs, and policy transfers. The cost of capital, defined as the pretax return that owners of capital require to maintain their investment in the system, is included in domestic costs (C); hence, profits (D) are excess profits-above-normal returns to operators of the activity. If private profits are negative ($D < 0$), operators are earning a subnormal rate of return and thus can be expected to exit from this activity unless something changes to increase profits to at least a normal level ($D = 0$). Alternatively, positive private profits ($D > 0$) are an indication of supernormal returns and should lead to future expansion of the system, unless the farming area can not be expanded or substitute crops are more privately profitable.

SOCIAL PROFITABILITY

The second row of the accounting matrix utilizes social prices. These valuations measure comparative advantage or efficiency in the agricultural commodity system. Efficient outcomes are achieved when an economy's resources are used in activities that create the highest levels of output and income. Social profits, H, are an efficiency measure because outputs, E, and inputs, $F + G$, are valued in prices that reflect scarcity values or social opportunity costs. Social profits, like the private analogue, are the difference between revenues and costs, all measured in social prices-$H = (E - F - G)$.

For outputs (E) and inputs (F) that are traded internationally, the appropriate social valuations are given by world prices-cif import prices for goods or services that are imported or fob export prices for exportables. World prices represent the government's choice to permit consumers and producers to import, export, or produce goods or services domestically; the social value of additional domestic output is thus the foreign exchange saved by reducing imports or earned by expanding exports (for each unit of production, the cif import or fob export price). Because of global output fluctuations or distorting policies abroad, the appropriate world prices might not be those that prevail during the base year chosen for the study. Instead, expected long-run values serve as social valuations for tradable outputs and inputs.

The services provided by domestic factors of production-labour, capital, and land - do not have world prices because the markets for these services are considered to be domestic. The social valuation of each factor service is found by estimation of the net income forgone because the factor is not employed in its best alternative use. This approach requires the commodity systems under analysis to be excluded from social factor price determination. For example, if land is planted to wheat, it cannot grow barley during the identical crop season; the social opportunity cost of the land for the wheat system is thus the net income lost because the land cannot produce barley. Similarly, the labour and capital used to produce wheat cannot simultaneously provide services elsewhere in agriculture or in other sectors of the economy. Their social opportunity costs are measured by the net income given up because alternative activities are deprived of the labour and capital services applied to wheat production.

The practice of social valuation of domestic factors begins with a distinction between mobile and fixed factors of production. Mobile factors, usually capital and labour, are factors that can move from agriculture to other sectors of the economy, such as industry, services, and energy. For mobile factors, prices are determined by aggregate supply and demand forces. Because alternative uses for these factors are available throughout the economy, the social values of capital and labour are determined at a national level, not solely within the agricultural sector. Actual wage rates for labour and rates of return to capital investment are therefore affected by a host of policies, some of which may distort factor prices directly. An enforced and binding minimum-wage law, for example, raises the market wage above what it would have been in the absence of policy and causes observed wages to be higher than the social opportunity cost of labour. But indirect effects can also be important. Distortions of output prices cause different activities to expand or contract, altering in turn the demand and prices of mobile domestic factors.

Fixed, or immobile, factors of production are the factors whose private or social opportunity costs are determined within a particular sector of the economy. The value of agricultural land, for example, is usually determined only by the land's worth in growing alternative crops. Because land is immobile, its value is not directly affected by events in the industrial and service sectors of the economy. But the social opportunity cost of farmland is sometimes difficult to estimate.

Within any agroclimatic zone, complete specialisation in the most profitable crop is rarely observed. Instead, farmers prefer rotations or intercropping systems that reduce risks of income losses from price variability, yield losses, and pest and disease infestation. Therefore, the social opportunity cost of the land is not accurately approximated by the net profitabilities of a single best alternative crop; instead, it is measured by some weighted average of the social profits accruing from the set of crops planted. Because the correct weights and social profits associated with each crop in the set are generally not known, it is convenient in assessing farming activities to reinterpret crop profits as rents to land and other fixed factors (for example, management and the ability to bear risk) per hectare of land used.

This reinterpretation includes private (and social) returns to land as parts of D (and H). Profitability per hectare is then interpreted as the ability of a farming activity to cover its long-run variable costs, in either private or social prices or as a return to fixed factors such as land, management skill, and water resources.

EFFECTS OF DIVERGENCES

The second identity of the accounting matrix concerns the differences between private and social valuations of revenues, costs, and profits. For each entry in the matrix-measured vertically-any divergence between the observed private (actual market) price and the estimated social (efficiency) price must be explained by the effects of policy or by the existence of market failures.

Table.2.2: Expanded Policy Analysis Matrix

	Revenues	Costs		Profits
		Tradable Inputs	Domestic Factors	
Private Prices	A	B	C	D
Social Prices	E	F	G	G
Diverges and efficient policy	I	J	K	L
Effects of market failures	M	N	O	P
Effects of distorting policy	Q	R	S	T
Effects of efficient policy	U	V	W	X

Notes: *Private profits,* D, equal A minus B minus C. *Social profits,* H, equal E minus F minus G. [3]Output transfers, 1, equal A minus E; they also equal M plus Q plus U. Input transfers, J, equal B minus F; they also equal N plus R plus V. Factor transfers, K, equal C minus G; they also equal O plus S plus W. Net transfers, L, equal D minus H; they also equal I minus J minus K; and they equal P plus T plus X.

This critical relationship follows directly from the definition of social prices. Social prices correct for the effects of distorting policies-policies that lead to an inefficient use of resources. These policies often are introduced because decision-makers are willing to accept some inefficiencies (and thus lower total income) in order to further non-efficiency objectives, such as the redistribution of income or the improvement of domestic food security. In this circumstance, assessing the tradeoffs between efficiency and non-efficiency objectives becomes a central part of policy analysis.

Some expended Policy Analysis are shown in Table Previous.

But not all policies distort the allocation of resources. Some policies are enacted expressly to improve efficiency by whenever monopolies or monopsonies (seller or buyer control over market prices), externalities (costs for which the imposer cannot be charged or benefits for which the provider cannot receive compensation), or factor market imperfections (inadequate development of institutions to provide competitive services and full information) prevent a market from creating an efficient allocation of products or factors. Hence, one needs to distinguish distorting policies, which cause losses of potential income, from efficient policies, which offset the effects of market failures and thus create greater income. Because efficient policies correct divergences, they reduce the differences between private and social valuations.

Interpretation of the effects of divergences can be clarified by the expansion of the PAM to include six rows. In this expanded PAM, each entry measuring the effects of divergences (I, J, K, and L) is disaggregated into three categories-market failures (fourth row), distorting policies (fifth row), and efficient policies (sixth row). The introduction of efficient policies to offset market failures would change the entries in the first and third rows. To bring about perfect efficiency, a government would introduce efficient policies to offset the effects of market failures and avoid distorting policies, thereby ensuring equality of private and social prices.

In the absence of market failure in the product markets, all divergences between private and social prices of tradable output and inputs are caused by distorting policy. Because the principles are identical for all tradable products, the matrix entries for revenues (tradable outputs)and tradable inputs can be considered together. Output transfers, I = (A - E), and input transfers, J = (B - F), arise from two kinds of policies that cause divergences between observed and world product prices: commodity-specific policies and exchange-rate policy.

Policies that apply to specific commodities include a wide range of taxes or subsidies and trade policy. For example, producer revenues per unit can be raised by producer subsidies (sometimes called deficiency payments in agriculture), tariffs or import quotas on outputs (which raise domestic prices), or domestic price supports enforced by government stockpiling (which require a complementary trade restriction for tradable products). Commodity-specific policies on inputs also affect private profitability. For example, per unit producer costs can be lowered by direct input subsidies or by subsidies on imported inputs.

Typically, PAM accounting is done in domestic currency, but world prices are quoted in foreign currency. Hence, a foreign exchange rate is needed to convert world prices into domestic equivalents. The social exchange rate may differ from observed exchange rates. Undervalued exchange rates reflect an excess supply of foreign exchange that is accumulating as excessive reserves and reducing potential income. Overvalued exchange rates correspond to conditions of excess demand; this demand results in extra foreign borrowing, excessive drawing down of exchange reserves, or rationing of foreign exchange among domestic users.

An overvalued exchange rate is an implicit tax on producers of tradable products because too little domestic currency is earned by exports or paid out for imports. In the absence of commodity policy, the world price of a tradable good determines its domestic price. When the exchange rate is overvalued, the domestic price is lower than its efficiency level and domestic producers are effectively taxed. Undervalued exchange rates exert the opposite effects. Correction for this distortion in PAM is done by conversion of world prices (E and F in the matrix) at the social exchange rate rather than at the official rate. Because exchange rates affect both product prices and factor prices, exchange-rate adjustments are limited to special circumstances-the appearance of multiple exchange-rate regimes or the government's failure to adjust the exchange rate enough to offset the effects of domestic inflation.

The social costs of domestic factors (G) reflect underlying supply and demand conditions in domestic factor markets. Factor prices are thus influenced by the prevailing set of macroeconomic and commodity price policies. In addition, the government can affect factor costs with tax or subsidy policies for one or more of the factors (capital, labour, or land) that create a divergence between private costs (C) and social costs (G). Finally, market imperfections, arising from imperfect information or underdeveloped institutions-which are often characteristic of developing country economies-further influence factor prices. If factor market imperfections exist along with distorting factor policy, both O and S and possibly W are positive components of K. The net transfer, L, thus combines the effects of distorting policy (I, J, and the S part of K) with those of factor market failures (the O part of K) and efficient policies to offset them (the W part of K).

The net transfer caused by policy and market failures (L in the matrix) is the sum of the separate effects from the product and factor markets, L = (I - J - K). (Positive entries in the two cost categories, J and K, represent negative transfers because they reduce private profits, whereas negative entries in J and K represent positive transfers; hence, J and K are subtracted from I, a positive transfer, in the calculation of the net transfer, L.) The net transfer from distorting policy is the sum of all factor, commodity, and exchange-rate policies (apart from efficient policies that offset market failures).

The net transfer can also be found by a comparison of private and social profits. These measures of the net transfer must by definition be identical in the double-entry accounting matrix, L = (I - J - K) - (D - H). Disaggregation of the total net transfer shows whether each distorting policy provides positive or negative transfers to the system. The PAM thus permits comparison of the effects of market failures and distorting policies for the entire set of commodity and macroprice (factor and exchange-rate) policies. This comparison can be made for the complete agricultural system and for each of its outputs and inputs.

COMPARISONS AMONG AGRICULTURAL SYSTEMS PRODUCING DIFFERENT OUTPUTS

The entries in PAM allow comparisons among agricultural systems that produce identical outputs, either within a single country or across two or more countries. In the accounting matrix, all measures are given as monetary units per physical unit of some commodity. If interest focuses solely on a comparison of one wheat system with another, for example, the matrix entries provide all information necessary for the analysis. Comparisons can be drawn readily by construction of PAM entries for two or more different systems that produce the same quality of wheat. (If necessary, premiums or discounts can be used to correct for quality differences.) Further comparisons can be made between the wheat systems in one country and those in other wheat-producing countries; social exchange rates, incorporating corrections for differential inflation not otherwise offset by exchange-rate changes, are used to convert the other countries' currencies into domestic currency.

Comparisons between wheat and barley-or apples and oranges are another story, however. To permit comparisons among systems producing different outputs, some common numeraire must be generated. One technique involves the expression of all values relative to a constraining domestic factor resource, such as land. A more common method uses ratios. Both the numerator and the denominator of each ratio are PAM entries defined in domestic currency units per physical unit of the commodity. Therefore, the ratio is a pure number free of any commodity or monetary designation.

PRIVATE PROFITABILITY

For comparisons of systems producing identical outputs, private profits, D = (A - B - C), indicate competitiveness under existing policies. Construction of a ratio is required to permit comparisons among systems producing different commodities. Direct inspection of the data for private profits is not sufficient. Profitability results are residuals and might have come from systems using very different levels of inputs to produce outputs with widely varying prices. This difficulty might.not be apparent in a wheat versus corn example, but it would arise in a comparison of a wheat system with one producing a high-value crop, such as strawberries. This ambiguity is inherent in comparisons of private profits of systems producing different commodities with differing capital intensities.

The problem is circumvented, by construction of a private cost ratio (PCR)-the ratio of domestic factor costs (C) to value added in private prices (A - B); that is, PCR = C/(A - B). Value added is the difference between the value of output and the costs of tradable inputs; it shows how much the system can afford to pay domestic factors (including a normal return to capital) and still remain competitive-that is, break even after earning normal profits, where (A - B - C) = D = 0. The entrepreneurs in the system prefer to earn excess profits (D > 0), and they can achieve this result if their private factor costs (C) are less than their value added in private prices (A - B). Thus they try to minimize the private cost ratio by holding down factor and tradable input costs in order to maximize excess profits.

SOCIAL PROFITABILITY

Social profits measure efficiency or comparative advantage. For a comparison of identical outputs, results can be taken directly from the second row of the PAM matrix-social profits equal social revenues less social costs, H

= (E - F - G). When social profits are negative, a system cannot survive without assistance from the government. Such systems waste scarce resources by producing at social costs that exceed the costs of importing. The choice is clear for efficiency-minded economic planners: enact new policies or remove existing ones to provide private incentives for systems that generate social profits, subject to non-efficiency objectives.

When systems producing different outputs are compared for relative efficiency, the domestic resource cost ratio (DRC), defined as G/(E - F), serves as a proxy measure for social profits. No new information beyond social revenues and costs is required to calculate a DRC. The DRC plays the same substitute role for social profits as does the PCR for private profits; in both instances, the ratio equals 1 if its analogous profitability measure equals 0. Minimizing the DRC is thus equivalent to maximizing social profits. In cross-commodity comparisons, DRC ratios replace social profit measures as indicators of relative degrees of efficiency.

POLICY TRANSFERS

Transfers are shown in the third row of the PAM. If market failures are unimportant, these transfers measure mainly the effects of distorting policy. Efficient systems earn excess profits without any help from the government, and subsidizing policy (L > 0) increases the final level of private profits. Because subsidizing policy permits inefficient systems to survive, the consequent waste of resources needs to be justified in terms of non-efficiency objectives.

Comparisons of the extent of policy transfers between two or more systems with different outputs also require the formation of ratios (for reasons analogous to those offered in the discussions of private and social profits). The nominal protection coefficient (NPC) is a ratio that contrasts the observed (private) commodity price with a comparable world (social) price. This ratio indicates the impact of policy (and of any market failures not corrected by efficient policy) that causes a divergence between the two prices. The NPC on tradable outputs (NPCO), defined as A/E, indicates the degree of output transfer; for example, an NPC of 1.10 shows that policies are increasing the market price to a level 10 percent higher than the world price. Similarly, the NPC on tradable inputs (NPCI), defined as B/F, shows the degree of tradable input transfer. An NPC on inputs of 0.80 shows that policies are reducing input costs; the average market prices for these inputs are only 80 percent of world prices.

The effective protection coefficient (EPC), another indicator of incentives, is the ratio of value added in private prices (A - B) to value added in world prices (E - F), or EPC = (A - B)/(E - F). This coefficient measures the degree of policy transfer from product market-output and tradable-input-policies. But, like the NPC, the EPC ignores the transfer effects of factor market policies. Hence, it is not a complete indicator of incentives.

An extension of the EPC to include factor transfers is the profitability coefficient (PC), the ratio of private and social profits or PC = (A - B - C)/(E - F - G), or D/H. The PC measures the incentive effects of all policies and thus serves as a proxy for the net policy transfer, since, L = (D - H). Its usefulness is restricted when private or social profits are negative, since, the signs of both entries must be known to allow clear interpretation.

A final incentive indicator is the subsidy ratio to producers (SRP), the net policy transfer as a proportion of total social revenues or SRP = L/E = (D - H)/E. The SRP shows the proportion of revenues in world prices that would be required if a single subsidy or tax were substituted for the entire set of commodity and macroeconomic policies. The SRP permits comparisons of the extent to which all policy subsidizes agricultural systems. The SRP measure can also be disaggregated into component transfers to show separately the effects of output, input, and factor policies.

DYNAMIC COMPARATIVE ADVANTAGE

The ability of an agricultural system to compete without distorting government policies can be strengthened or eroded by changes in economic conditions. Dynamic comparative advantage refers to shifts in a system's competitiveness that occur over time because of changes in three categories of economic parameters - long-run world prices of tradable outputs and inputs, social opportunity costs of domestic factors of production (labour, capital, and land), and production technologies used in farming or marketing. Together, these three parameters determine social profitability and comparative advantage.

The appropriate world prices for measuring efficiency or comparative advantage are long-run equilibrium levels that approximate best guesses of expected future prices. If the country's decisions to buy or sell on world markets will not have any measurable effect on world price levels, those price levels can be considered exogenous and, once arrived at, can be taken as given for domestic agricultural systems.

The world prices are the correct indicators of social valuation of tradable commodities even if a country's decisions to buy or sell internationally do affect the world price of a good. When a large country has market power, however, the analyst needs to take into account the impact of that country's trading decisions on world prices.

In the absence of knowledge of future prices, most analysts project constant long-run real prices rather than fluctuating prices. If new information results in changes in the constant price guess or in the projection of continually increasing or decreasing future prices, these changes can be incorporated easily into the PAM. Separate PAMs can be constructed for each year, and each can have different assumed world prices.

Costs of factor services in any country can be expected to change over time. But cyclical variations in the real wage and the real return to capital, associated with swings in macroeconomic policy, are not the primary focus of the PAM method. Instead, interest centers on long-run trends in the costs of labour, capital, and land. As economies grow, real wages typically rise, both in absolute terms and relative to real costs of capital and land. For agricultural systems, changes in the social opportunity costs of labour and of capital depend on changes in the national environment for investment and growth. Land rental rates are endogenous to agriculture but will be constrained by changes in world prices and in real wage and interest rates, because payments to land and other permanently fixed factors come out of profits. Analysis of projected comparative advantage therefore includes both the future pressures that changing real factor prices might exert on agricultural systems and the influences of likely world prices for tradable outputs and inputs. The results identify systems that can readily expand and those that will have to contract or change in order to survive.

Changes over time in factor and commodity prices can also influence agricultural technologies. Farmers and researchers innovate, often by finding new ways of using less of factors that are relatively expensive (usually labour) and more of other inputs. Successful technological change permits commodities to be produced with reduced costs of one or more inputs. Empirical analysis of intra-system change can be done with partial budgeting, a technique in which individual cost-saving or revenue-increasing changes can be analyzed within the PAM for the initial system.

CONCLUDING COMMENTS

The central purpose of PAM analysis is to measure the impact of government policy on the private profitability of agricultural systems and on the efficiency of resource use. Private profitability and competitiveness are likely to be uppermost in the minds of those concerned specifically with agricultural incomes. Social profitability and efficiency are often emphasized by economic planners whose concern is the allocation of resources among sectors and the growth of aggregate income in the economy. Both sets of issues ultimately focus on the incentive effects of policy-part of the difference between private and social profitability-and on how policy incentives might be altered. Through evaluation of private and social revenues and costs, the PAM method is designed to illuminate these related issues of agricultural policy analysis. The approach is particularly well suited to empirical analysis of agricultural price policy and farm incomes, public investment policy and efficiency, and agricultural research policy and technological change.

The PAM approach to policy evaluation advocates a disaggregated view of efficiency effects (as measured by social profitability) and of non-efficiency effects. The analyst can do much in describing the contributions of a particular system to non-efficiency objectives and in quantifying implications for efficiency (aggregate income gains or losses). But it is left to the discretion of each policy-maker to determine whether tradeoffs between efficiency and non-efficiency objectives merit changes in policy or maintenance of incentives to particular systems.

In other approaches to policy analysis, it is desirable to aggregate measures of efficiency and non-efficiency effects into a single measure. Income distribution concerns, for example, can be introduced into the social cost estimates by weighting (with a value less than 1) of the efficiency-determined value of unskilled labour wages. Concerns for food self-sufficiency can be introduced by the addition of a premium to the world market value of output. If these weights were incorporated in the calculations of social profitability, the policy-makers' decision would become automatic and predetermined: encourage all systems with positive social profitability and discourage all systems with negative profitability.

The disadvantage of the aggregate approach lies in its tendency to lump together high-quality information (observable data on prices and input-output relationships) with relatively poor-quality information (implicit weights of society or of policy-makers regarding various prices of inputs and outputs). Moreover, attempts to quantify implicit policy weights presume the existence of some dictatorial policy-maker who speaks on behalf of society. Policy-making rarely occurs in such an environment. Policies are the outcomes of negotiated conflict between interest groups both within and outside the government. Quantitative studies provide improved information and thus increase the probability of good policy decisions. But these decisions, and the tradeoffs implied between efficiency and non-efficiency objectives, are the outcomes of debate based on this information, not inputs into the collection of information.

8

Agricultural Productivity in Development Countries

AGRICULTURAL PRODUCTIVITY

The measures of agricultural productivity and their rates of change, by region and country, based on data from the World Bank (1998) and Lusigi and Thirtle (1997). It is important to note that the value of agricultural output used in generating these statistics is itself difficult to estimate because a large share of agricultural output is either consumed within the household that produces it, or is exchanged for commodities other than money (World Bank, 1998). The value of output is often estimated indirectly using a combination of methods, including reliance on estimates of yields and area under cultivation. To the extent that such attempts underestimate the true magnitude of agricultural output, estimates of agricultural productivity will themselves be biased downwards.

Output per unit of land, or crop yield, is commonly used by agricultural scientists to assess the success of new production practices. Land productivity is also used by national policy-makers to assess agricultural production for meeting national food security needs. Output per agricultural worker, on the other hand, may be a more important indicator of rural standards of living and welfare. Recognizing that food may be acquired through exchange as well as production, income becomes an important determinant of access to food and thus of food security. As a result, labour productivity may be particularly important as an indication of the ability of agricultural workers to acquire sufficient food, whether or not they produce food themselves. Land productivity averaged US$68 per hectare of agricultural land (measured as the sum of arable land, permanent cropland and permanent pasture) for SSA as a whole in 1993, compared with $519 in South Asia and $116 in Latin America and the Caribbean. Values ranged from $5-10 per hectare in the drier countries of Southern Africa and the Sahel to $200 per hectare and more in the East African highland countries and tropical West Africa. For SSA as a whole, land productivity grew at an average rate of 1.9 percent per year between 1980 and 1993, with slow to moderate growth in most countries. Land productivity grew most rapidly in the Sahelian countries and West Africa, and more slowly in Eastern and Southern Africa. These patterns in land productivity levels and growth.

Labour productivity averaged $392 per agricultural worker for SSA as a whole in 1995, this compares with $383 in South Asia and $2 292 in Latin America and the Caribbean. Values ranged from $100-$200 per worker in many countries in Eastern and Southern Africa and the Sahel to more than $500 per worker in parts of West and Central Africa. Labour productivity declined at an average rate of 1.0 percent per year for SSA as a whole between 1980 and 1995, with modest growth in West Africa and declines in Eastern and Southern Africa. In contrast to the pattern evident for land productivity, the most rapid growth in labour productivity appears to be taking place in those countries that already have the highest levels of labour productivity. Perhaps this reflects the pull of off-farm employment opportunities, suggesting that disparities in labour productivity across countries may increase over time.

Low (or declining) labour productivity is consistent with high (or growing) land productivity in the context of a large (or expanding) agricultural labour force. Such patterns are evident in the agricultural labour/land ratios. The labour/land ratio is generally high in East Africa and low in Central and Southern Africa and the Sahel.

Land and labour productivity measures are both incomplete indicators of agricultural productivity, since they measure the productivity of only a single factor of production, and may well move in opposite directions. In an effort to address this problem, economists estimate total factor productivity (TFP), which measures changes in agricultural output relative to changes in an aggregated index of multiple inputs. TFP growth reflects factors such as technical change or improvements in infrastructure or research. It can also reflect failure to include or measure inputs such as depletion of soil or other natural resources. For Africa as a whole the TFP index estimated by Lusigi and Thirtle averaged 0.8. This indicates that on average, the productivity of the set of inputs measured was 0.8 times as high in 1991 as it was in the most efficient countries in 1961, up from an average of 0.7 in 1961. Levels are generally mixed, with the highest estimates in Uganda and Burundi, suggesting the importance of land quality, availability of water, and labour supply in driving agricultural TFP. Uganda also has one of the highest rates of growth in TFP, averaging 7.8 percent per year since 1961. For Africa as a whole, TFP grew at an average rate of 1.3 percent per year over the period.

FACTORS AFFECTING PRODUCTIVITY GROWTH

In explaining productivity growth, economists originally limited themselves to the role of conventional inputs such as land, labour, physical capital, water and chemical inputs. However, the failure to explain productivity growth adequately led them to examine the role of human capital and public goods, such as education, agricultural research and extension and publicly provided infrastructure (Griliches, 1963; Mankiw, Romer and Weil, 1992; Nelson, 1964 and 1981; Solow, 1957). Public policies that have a strong link to agricultural productivity such as policy reforms were also examined.

The rationale for considering research is the belief that investments in research result in increases in the stock of knowledge, which, in turn, either facilitate the use of existing knowledge or generate new technology. Technological advances, whether resulting from changes in input quality or how inputs are combined, lead to productivity gains. Education, training and extension also increase productivity by increasing people's knowledge and skill base, which are essential for technology adoption and efficient use of inputs. Public infrastructure, on the other hand, increases productivity by facilitating the exchange of goods and services.

Technological Change

Technological change is recognized by many as one of the most important sources of productivity growth. It refers to the changes in the production process that come about from the application of innovation and newly acquired scientific knowledge and technical and management skills. Technological change increases agricultural productivity either by shifting the production frontier upward so that more measured output can be produced with the same amount of inputs or by moving closer to the production frontier so that the same amount of output can be produced with a smaller amount of inputs. Better organizational and management skills not only improve input-output combinations but enable producers to respond more quickly to changing market circumstances.

While generation of new technology or knowledge comes from investments in research and development, adoption of technology involves investments by the potential users in both physical and human capital. Therefore, adoption of technology depends principally on their applicability and expected returns of the innovation. However, there may be a long lag between development, adoption and productivity gains. Chavas and Cox (1992) found the lag to be up to 15 years between making an investment in research and having an effect on productivity. However, after taking effect, the benefits from an innovation may persist for thirty years or more.

The lag between generation of new technology and its widespread adoption by farmers has important policy implications. First, the adverse effects of reduced public funding to agricultural research and extension on productivity may be under-estimated if the lagged effects are not accounted for. Secondly, the complementarity between research and extension should be taken into account. The former helps the development of new technology, while the latter helps speed up the rate of diffusion and adoption of new technology. Extension can be done more effectively by identifying factors that contribute to technology adoption. As an example, innovators in a farming community can be identified and targeted for extension services. Since better-educated farmers are found to be more likely to adopt new technology, human capital is a pre-condition for technology adoption and hence productivity growth. Further, if adoption of new technology requires additional investments, lack of access to credit and additional inputs may prevent or slow down technology adoption. Finally, because potential users of new technology often differ in the agronomic-ecological conditions in which they operate, new technology may require adaptive research before it can be transferred successfully to different locations. These impediments to technology adoption mean careful planning and provision of necessary infrastructure are essential to capture the full benefits of new technology.

Agricultural Research and Extension

Many researchers have explored the roles of research and extension in promoting agricultural growth. Rosegrant and Evenson (1992) found that in South Asia, public research accounted for 30 percent of the output growth, and extension for about 25 percent, with corresponding rates of return being 63 percent and 52 percent, respectively. Pray and Evenson's (1991) survey of Asia found the rates of return to national research investment ranged from 19 to 218 percent, returns to national extension investment from 15 to 215 percent, and returns to international research investment from 68 to 108 percent.

Evenson and McKinsey (1991) found that public investment in research accounted for over half of the output growth in India and extension contributed about one-third. The calculated internal rates of return were 218 percent for public research and 177 percent for extension. However, they found that little output growth was attributable to infrastructure. Jamison and Lau (1982) also found that physical capital had little impact on production or profits, as compared to farmer's education and extension services.

Fan (1996) found that public research expenditures accounted for about 20 percent of total production growth in Chinese agriculture during the period 1965 to 1994. The annual rates of return to agricultural research investment in China ranged from 44 percent to 83 percent.

Fan (1996) concluded that the rapid growth in agricultural output in China during the 1980s and 1990s was the result of public investments in R&D as well as the institutional and market reforms that began in 1979. He concluded that increases in agricultural research were justifiable; not only did they stimulate additional output growth, but the rate of return to agricultural research was much higher than commercial interest rates. Despite the high rates of returns from public research investments, agricultural research intensity (ARI), measured as a percentage of Chinese agricultural GDP, was found to have declined from 0.56 percent for the period 1958-1965 to 0.43, 0.44, 0.39 and 0.40 percent, respectively, for 1966-1976, 1977-1985, 1986-1990 and 1991-1993 (Fan, 1996). Lin (1998) reported that, as part of the overall market reform, the Chinese Government had reduced its fiscal appropriation for agricultural research, shifting funding from institutional supports to competitive grants and cost recovery. As such, it can be expected that an increasing proportion of research activities will move from the public to the private domain. Other studies on output growth have also shown a high payoff from agricultural research and extension. The results indicate that the rate of return on research, in most cases, ranged from 15 to 50 percent for both developed and developing countries, but some estimates were as high as 218 percent. The wide disparity among the estimates raises questions regarding the sensitivity of these estimates to the commodity of interest and the use of different time periods and methodologies.

Estimates for Asian countries appear to be higher and show a much wider variation than those of studies in the United States. This could be due to the diverse nature of Asian agriculture, which differs from country to country in economic, social and agronomic-ecological conditions. Because of inconsistency in the data and methodology used, it is not possible to make direct comparisons across countries or over time. Nevertheless, the general conclusion that R&D yields relatively high returns seems indisputable.

PRODUCTIVITY GROWTH IN AGRICULTURE

Economists originally limited themselves to examining the roles of labour and physical capital in economic growth. The failure to adequately explain growth led them to examine the roles of other factors and to develop endogenous growth theory. Investment in infrastructure has been cited as an important source of growth in agriculture. However, Ferreira and Khatami (1996) claim that economic literature has not reached a consensus on the direction of causality between infrastructure and development. Nor can investment be viewed in isolation of policy reform which has been shown to be a vital stimulus of production; as have institutions. Public investment in forms of human capital: education, extension, training and technology research have also been shown to increase productivity.

Nelson (1964 and 1981) recognized that there are important interactions between capital formation, labour allocation, technical progress and productivity. This calls into question whether the growth due to physical capital can be separated from growth attributed to other inputs. Unless a production technology is a fixed Leontief process, there is always some degree of substitutability among categories of inputs. However, since inputs are not perfect substitutes, the lack of adequate investment can slow down production growth. Estimates of the elasticity of substitution in agriculture between hired labour and capital equipment vary from 0.32 in the short run to 1.78 percent in the long run.

Most measures of TFP incorporate inputs and physical capital, leaving human and social capital, technology, institutions, infrastructure and policy to "explain" growth in TFP. Social and human capital are the on-farm human elements that mediate how policy, technology, institutions and infrastructure affect input and physical capital use. Human capital directly affects whether and how technology will be adopted. Technology choice in turn, affects the inputs and physical capital used. That is, technology is embodied in the types of inputs and how they are used. Social capital affects access to physical capital (*e.g.* land directly or through land titling and loans) and variable inputs (*e.g.* through credit or cooperatives).

In general, researchers have estimated TFP and then focused on how one or several of these factors might be driving its growth. Usually, they have done so using the change in TFP as a dependent variable in a regression with explanatory variables that represent measures of technology, human capital and policy (which are not easily quantifiable or assignable in constructing the production indices). In the following sections, policy is divided between budgetary policies that affect investment in R&D and infrastructure, political and economic policies and political stability.

HUMAN CAPITAL INFLUENCES AGRICULTURAL PRODUCTIVITY

Human capital directly influences agricultural productivity by affecting the way in which inputs are used and combined by farmers. Improvements in human capital affect acquisition, assimilation and implementation of information and technology. Human capital also affects one's ability to adapt technology to a particular situation or to changing needs.

Schultz's (1963) classic work attributed between 21 to 23 percent of the growth in U.S. income, between 1929 and 1957, to education of the labour force. Contemporaneously, Griliches (1963) focused on minimizing the unexplained portion of growth in U.S. agriculture by adjusting labour for quality, using education. When he included research and extension expenditure as an input to production, he found that virtually all the "unexplained" growth could be explained by economies of scale, R&D and labour quality changes. Romer (1986) and Lucas (1988) provide theoretical grounds for human capital being the driving force behind economic growth.

Jamison and Lau (1982) explored the role of farmer education and extension on farm efficiency. They found that farmer education and extension were not only important to enhancing production on Thai, Korean and Malaysian farms, but that there was an interaction effect between education and extension. In contrast, they found physical capital had an insignificant impact on production and profits. On the other hand, some researchers are finding evidence that returns to education are low, especially for those who stay in agriculture. In their summary of the findings on the determinants of rural poverty for six country studies based on econometrically estimated income equations, Lopez and Valdes (2000) conclude that the return to education in farming is surprisingly small in most cases. An increase in one year in the average level of schooling raises per capita annual income of the family by less than US$ 20 per person in most cases. The main contribution of education in rural areas appears to be to prepare young people to emigrate to urban areas and towns.

Using an econometric approach, Nehru and Dhareshwar (1994) examined sources of TFP growth in 83 industrial and developing countries for the period 1960-1990. They found that human capital formation was three to four times more important than raw labour in explaining output growth. Using human capital as a separate variable, they found that the countries with the fastest growing economies have based their growth on factor accumulation (human capital, labour and physical capital), not growth in efficiency or technology.

Research and Technology Transfer

Research increases the set of available technologies, hence agricultural R&D expenditures are used as a proxy for agricultural technological change. However, the development of technology does not always result in its adoption. In some cases this may be because the technology being developed is not appropriate, that is, it does not meet the needs of agricultural producers. Hence, researchers focus on public expenditure as an explanatory variable in TFP growth. Additionally public research has been shown to lead private research.

Several caveats arise in focusing on public R&D to explain growth in agricultural TFP. Public R&D expenditures are used as proxy for R&D results, yet there is not an exact correspondence between expenditures and technology. Even when technology is produced, researchers may have different goals than farmers, *e.g.* yield maximization rather than profit maximization or risk minimization or improvement in commercial crops rather than staple crops. Additionally, when an appropriate technology does result, the process of technology adoption in agriculture is widely recognized as one that occurs over many years in which some adopt quickly and others wait for extension or the results of their neighbours to convince them to adopt.

Bearing this in mind, researchers have found that public investment in developing and extending agricultural technology is justified by the high rates of return to such investment. In a survey of studies on Asia, Pray and Evenson (1991) found rates of return to national research investment from 19 to 218 percent, returns to national extension investment from 15 to 215 percent and returns to international research investment of 68 to 108 percent. A report of the Taskforce on Research Innovations for Productivity and Sustainability indicated that the returns to research, though variable, were always high, from 22 to 191 percent. Using an index number approach to calculate TFP for several crops in India, Rosegrant and Evenson (1992) and Evenson and McKinsey (1991) used econometric analysis to identify sources of growth in TFP. Rosegrant and Evenson (1992) found that public research accounted for 30 percent of growth and extension for about 25 percent, with rates of return for each respectively of 63 percent and 52 percent. Evenson and McKinsey (1991) found that public investment in India in research accounts for over half of growth, while extension contributes about one-third and infrastructure accounted for very little growth. They calculated internal rates of return of 218 percent for public research, 177 percent for public extension and 95 percent for private research expenditures in India.

Block (1994) compares econometric estimates of TFP for Sub-Saharan Africa between 1963 and 1988. He uses three different methods of aggregating agricultural output: official exchange rates, purchasing power parity and wheat units. He finds that one-third of the growth in agricultural TFP in Sub-Saharan Africa is due to research expenditures. In India, Rao and Hanumantha (1994) attribute continued growth in agriculture, despite a sharp decline in physical capital formation, to better utilization of existing infrastructure, fertilizer and high yielding varieties.

While the returns to research are high, the technology is not always adopted. For example, high yield varieties (HYV) of wheat and rice have been introduced on less than one-third of the 423 million hectares planted to cereal grains in the Third World. Specifically, in Asia and the Middle East 36 percent of the grain area was HYV, 22 percent in Latin America and one percent in Africa (Wolf, 1987). This implies there is much potential for increasing agricultural productivity using existing technology. However, the use of HYV requires increased use of fertilizer, but external debt in Latin America and poverty and inadequate water supply in Africa have made fertilizer use and hence HYV unprofitable. Jahnke, Kirschke and Lagemann (n.d.) also attributed low adoption of HYV in Africa to lack of appropriate technology development and few extension services directed to women. Additionally, nontraditional crops have rarely been the focus of improved varieties or technology and potential exists to develop them to increase agricultural production.

Public Investment and Policy

Public policy and budgetary decisions regarding infrastructure also have a profound effect on agricultural production. The financing aspects of public R&D and human capital development, but both physical and institutional infrastructure affect the development and transfer of technology. For example, irrigation systems and roads may be required to make a technology profitable to implement. Reforms in pricing policy or the marketing system may be needed to provide incentives. A serious conflict arises with structural adjustment reforms. Budget cuts in public services often accompany market reforms. While fiscal restraint may be required to stabilize the economy in the short run, cuts in human capital development, public R&D, and infrastructure have a detrimental long-term effect on productivity growth. Policy makers need to choose carefully to mitigate the deleterious impacts of budget cuts on future growth.

Using an econometric approach, Jayne *et al.* (1994) demonstrated the complementarity of public policies and public investments in facilitating the use of new technology. They point to the sharp decline in public investments and growth in Zimbabwe during the 1980s. Pal (1985) underscores the complementarity of public policy towards investment in irrigation technology and private variable input use.

The importance of policy reform is increasingly viewed as fundamental for agricultural productivity gains. Liberalizing markets so prices can send proper signals to producers is the fundamental objective of structural adjustment programmes in developing countries and policy reform in economies in transition. Assigning property rights is viewed as a means of promoting development through the efficient and responsible use of resources (North, 1994) and therefore underlies the distribution of capital in economies in transition, land reform and most land policy. Block (1994) discusses the complementarity of economic reform and technical change, but cautions that policy reform offers a one-time effect.

An example of the relation between policy reform and productivity is the implementation of China's "responsibility system" (RS) in 1980-81, which linked productivity to material reward, resulted in increased crop yields "for every major crop". McMillan, Whalley and Zhu (1989) calculated that in response to the RS and price reforms, output in the Chinese agricultural sector increased by over 61 percent between 1978 and 1984. They attribute 78 percent of the increase to the RS and 22 percent to higher prices for crops. They calculate the RS increased productivity in agriculture by 32 percent. Lin (1992) calculated that 42 to 47 percent of the growth in agricultural output was attributable to the RS during the same period.

In another example, price reforms in Egypt implemented in 1986 resulted in increased wheat and maize yields from 1987 to 1993. Rice production increased by 62 percent, while yields increased by 42 percent. Bevan, Collier and Gunning (1993) contrast the performance of agriculture in Kenya and Tanzania. In Kenya where there was little intervention production of food and cash crops increased by 4.6 and 5.5 percent per annum, respectively. In Tanzania, where policies controlled prices and taxed export crops, agricultural production stagnated until policy reforms were instituted in 1984. Using an econometric approach to estimate TFP for the United States dairy industry

1972-1992, Lachaal (1994) examined how protectionist policies in the form of direct subsidies to agriculture reduced productivity growth in the United States dairy industry. Lachaal showed that government subsidies encouraged using materials at the expense of feed and raised the cost of production by 1.8 percent for each 10 percent increase in subsidy. The subsidy policy was the source of technical inefficiency, creating biases that distorted factor usage.

Political Stability and Conflict

Another aspect of policy that can influence or hinder agricultural production is the political situation. In a study of the productivity growth of 83 industrial and developing countries between 1960 and 1990, Nehru and Dhareshwar (1994) found that the economies that perform the worst are those involved in wars (particularly civil wars) and those that have the most price distorting policies. They explore a variety of policy variables and find that apart from political stability and the initial endowments of a county, virtually no other policy variable is associated with growth.

The World Food Summit Plan of Action items 2 and 3 (1996b) recognize the role of government in providing an environment conducive to investment, through guarantee of rights and law as well as policies encouraging investment. Corruption is the extreme case where law enforcement breaks down and incentives are lacking. While long-standing institutionalized bribery can be seen as simply an added cost of doing business, pervasive corruption and violence increase risk and result in capital flight, disinvestment and jeopardize assistance.

PRODUCTIVITY GROWTH IN GLOBAL AGRICULTURE SHIFTING TO DEVELOPING COUNTRIES

Improving agricultural productivity has been the world's primary defence against a recurring Malthusian crisis—where needs of a growing population outstrip the ability of humankind to supply food. Over the last half-century, world population doubled while food supply tripled, even as land under cultivation grew by only 12 per cent. It is by raising productivity, or getting more output from existing resources, that has been driving growth in global agriculture, and what has proven Malthus wrong. In fact, at the global level, the long-run trend since, at least 1900 has been one of increasing food abundance—in inflation-adjusted dollars, food prices fell by an average of 1 per cent per year over the course of the 20th Century.

But since, around 2001, real food prices have been rising. While demand-side factors—continued population growth, greater per capita consumption of meat, diversion of crop commodities for biofuel—and weather-induced production shocks—like the 2012 drought in North America—are generally thought to be the major forces behind these rising prices, the persistence of high prices has renewed concerns about whether agriculture is facing new constraints on supply growth. In particular, if agricultural productivity is reaching its limit, it may harbinger even further increases in commodity prices, rising food scarcity, and increased competition for land, water and energy resources. In fact, for major cereal grains like wheat and rice, average rates of yield growth have slowed from about 2 per cent per year in the 1970s and 1980s to about 1 per cent per year since, 1990. There is also evidence that growth in agricultural total factor productivity (a broad measure of sector-wide productivity) may have slowed down in some developed countries.

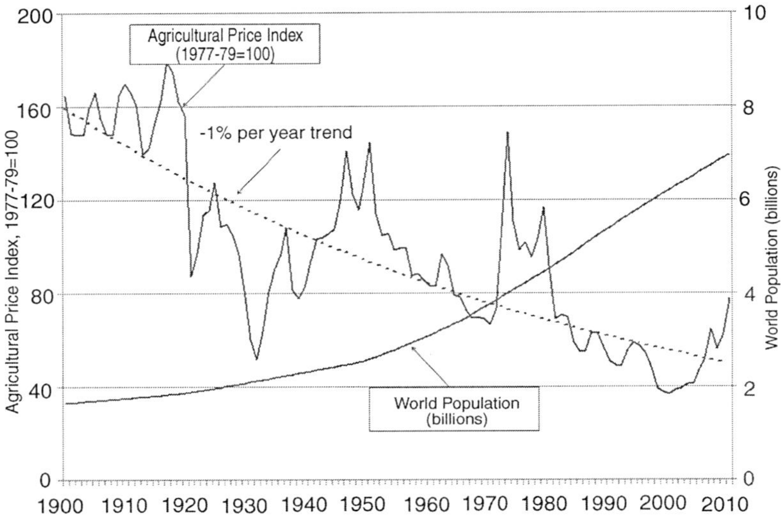

Fig. Real Agricultural prices have fallen dramatically since, 1900, even as world population soared

In recent decades world agriculture has undergone some fundamental changes. One has been that many developing countries have greatly expanded their capacities in agricultural research and innovation, providing many new improvements in farming technologies and practices. Complementing this expanded research capacity have been institutional and policy reforms, improvements in farmer education and health, and investments in rural infrastructure, all of which help create an environment in which new farm technologies and practices are adopted more rapidly. A second major development has been the shifting location and composition of global agricultural production. With slower agricultural growth in developed countries and—following the collapse of the USSR—a significant contraction in agricultural output in former Soviet bloc countries, developing countries now account for about two-thirds of global agricultural output—up from 42 per cent in 1961.

Further, as rising incomes in developing countries lead to changes in the kinds of foods consumers demand, the share of staple food commodities in world agriculture has declined.

The share of global agricultural output contributed by cereal grains and root and tuber crops fell from 30 per cent in the 1960s to 24 per cent in 2010, while the gross value share in fruits and vegetables increased from 18 per cent to 22 per cent, and the share in oil crops rose from under 5 per cent to nearly 8 per cent. Meanwhile, the crop-animal shares of gross agricultural output have remained stable, at roughly 63 per cent in crops. These developments suggest that the global picture of agricultural productivity depends heavily on how it evolves in developing countries and amongst a broad set of commodities, not just food staples.

The 15 case studies investigate agricultural productivity growth and its drivers across a broad swath of the globe. Included are developed countries (the United States, Western Europe, Canada, Australia and South Africa), developing countries and regions (Brazil, China, India, Indonesia, Thailand and Sub-Saharan Africa), and transition countries (Eastern Europe and the former Soviet Union). Taken together, the case studies point towards robust but highly uneven productivity growth in global agriculture.

PRODUCTIVITY GROWTH IN AGRICULTURE: AN INTERNATIONAL PERSPECTIVE

International comparisons of productivity usually begin by comparing agricultural land and labour productivity. Over the past 50 years, the highest levels of agricultural output per worker and per acre of agricultural land have been consistently achieved by industrialized nations. Currently, the world's highest yields—gross output of crops and livestock per hectare of land—are found in the developed countries of northeast Asia (Japan, South Korea and Taiwan), while the highest output per agricultural worker is in North America (the United States and Canada) and Oceania (Australia and New Zealand). Developing countries lag far behind, but in recent decades they have been slowly catching up.

Developing countries are currently at levels achieved by industrialized nations in the 1960s, suggesting that large global productivity gaps persist among countries.

Much of the gains in land and labour productivity came from more intensive use of other inputs—such as fertilizers, machinery, energy and irrigation. A broader concept of agricultural productivity is total factor productivity (TFP), which is the ratio between total outputs of crops and livestock to total inputs—an aggregation of all of the land, labour, capital and materials used in production.

Growth in TFP is then the difference between aggregate output growth and input growth. In practice, measuring TFP is a data intensive exercise requiring detailed historical statistics on the quantities and prices of outputs and inputs, and the use of appropriate index methods to account for input and output substitution possibilities as relative prices change. Moreover, empirical measures of agricultural TFP rarely account for quality improvements in inputs or changes in natural resource stocks—such as biodiversity, water quantity and quality, and greenhouse gas emissions—that result from agricultural activity.

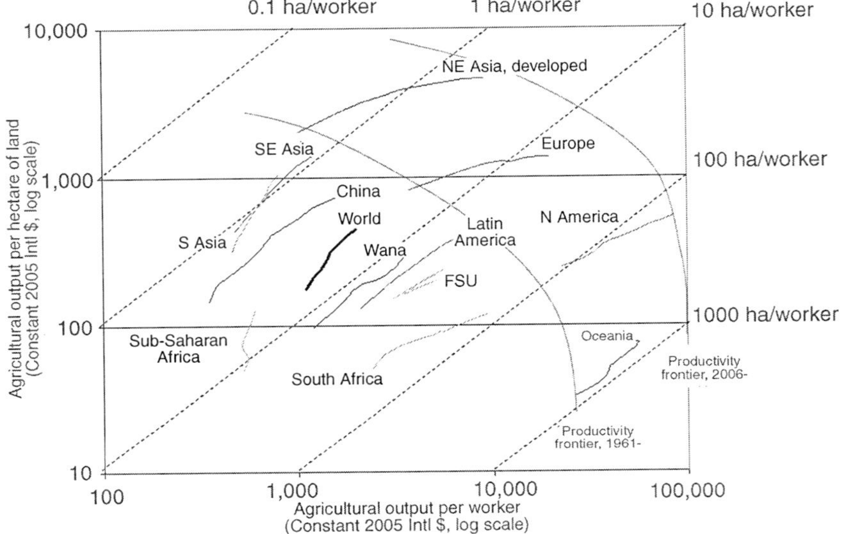

Fig. Agricultural land and labour productivity has steadily improved, but developing countries lag decades behind developed countries.

However, while agricultural production is generally thought to draw down these natural resource stocks, what this implies for TFP is not immediately evident. While the evidence is far from complete, some studies suggest that productivity improvement in agriculture has significantly reduced negative environmental externalities from agriculture—and thus conventional measures underestimate growth in TFP—through averting forest-to-cropland conversion and reducing greenhouse gas emissions per ton of meat and milk output from ruminant livestock. On the other hand, the expansion of large-scale, confined hog feeding operations in the United States has been a significant part of TFP growth in this industry, but the concentration of wastes from these operations poses greater risks for water and air pollution. A number of productivity measures for the global agricultural economy as a whole, by decade since, 1961, including TFP. Output growth has remained remarkably constant over these decades, averaging 2.7 per cent per year in the 1960s and between 2.1 per cent to 2.5 per cent per year every decade since, then.

Growth in agricultural yield—total output per hectare of agricultural land—has mimicked the trends in output growth, remaining fairly steady over the past 50 years around an average of 2.1 per cent annually. The growth rate in cereal yield per area harvested, however, has shown signs of slowing after 1990.

One reason the decline in area harvested yield is not reflected in total agricultural yield is because of land-use intensification. More double cropping and less idled or fallow cropland means more harvests in a year, on average, per hectare of cropland.

This can keep cropland productivity rising, even if yield per harvest is stagnant. The decline in cereal yield growth is also being offset by other productivity improvements in the sector—such as rising yield growth in other crop commodities, and more meat and milk output per animal, acre of pasture, and pound of feed—to keep total output per hectare of agricultural land rising at historical rates.

Table. Productivity Indicators for World Agriculture Show Robust Growth Despite Slowing Rates of Cereal Yield Improvement.

Period	Gross Output	Total Input	Total Factor Productivity	Output per Worker	Output Per Area of Cropland	Cereal Yield Per area Harvested
			Average Growth Rate (% Per Year)			
1961-70	2.74	2.55	0.18	1.13	2.45	2.88
1971-80	2.3	1.7	0.6	1.58	2.09	2.08
1981-90	2.12	1.5	0.62	0.62	1.75	1.88
1991-00	2.21	0.55	1.65	2	2.16	1.57
2001-09	2.49	0.65	1.84	2.8	2.64	1.8
1971-90	2.25	1.53	0.72	1.11	1.97	2.25
1991-09	2.29	0.7	1.59	1.97	2.27	1.42
1961-09	2.23	1.28	0.95	1.19	2	1.99

Gross agricultural output is from FAO and is measured in constant 2005 international dollars. Total input is the aggregation of agricultural land, labour, capital and material inputs. Growth in TFP is the difference between output growth and total input growth. Output per worker is gross agricultural output divided by number of economically active adults working in agriculture. Output per hectare is gross agricultural output divided by total cropland plus permanent pasture. Cereal yield is global output of maize, rice and wheat (in metric tons) divided by the area harvested of these crops. The average annual growth rate in a series (call it Y) is found by regressing its log value against time, *i.e.,,* the estimate of parameter β in $\ln(Y) = A + \beta t$.

The TFP measure of productivity change captures a broader set of productivity improvements—including those that save agricultural resources other than land. It does not count as productivity growth simple substitution between inputs—fertilizer for land or machinery for labour, for example, if this doesn't save costs.

Thus, TFP provides a better measure of the underlying rate of technical change. And according to our estimates, the average rate of technical change (TFP) in global agriculture rose significantly over the past half-century. The growth rate in aggregate inputs used in agriculture, meanwhile, fell steadily. Over the past five decades, the source of growth in the global agriculture output has shifted dramatically from being primarily resource-driven to primarily productivity-driven.

The sources of growth in global agricultural output into contributions from:

- Growth in land and water (irrigation),
- Intensification of other inputs per unit of land, and
- TFP. Over these five decades, total inputs grew at about 60 per cent as fast as gross agricultural output, implying that improvement in TFP accounted for about 40 per cent of total output growth. Moreover, TFP's contribution to output growth grew over time, and by the most recent decade (2001-2009), TFP accounted for about three-quarters of the growth in global agricultural production. The rate of expansion in natural resources used—land and water—has slowed slightly over time, while the rate of growth in input intensification—the amount of labour, capital and materials per hectare of land—has fallen sharply. The source of increase in agricultural yield—output per hectare of agricultural land—has shifted markedly from input intensification to improvement in TFP.

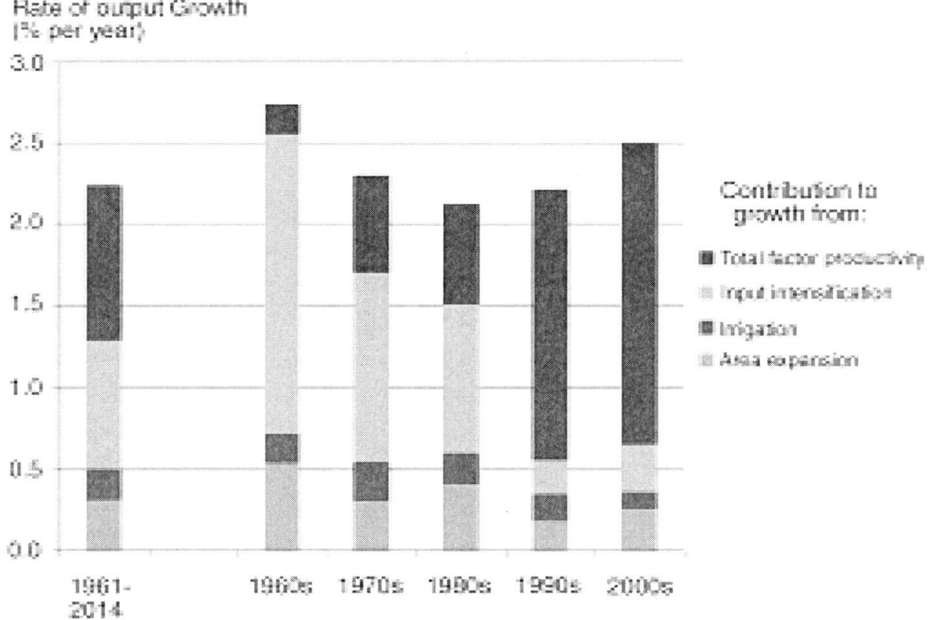

Fig. Total factor productivity has replaced resource expansion and input intensification as the primary source of agricultural growth in world agriculture

AGRICULTURAL PRODUCTIVITY IS GROWING

The estimates of global agricultural output and TFP growth are disaggregated among global regions. The trends are hardly uniform, but three general patterns are evident:

Table. Agricultural Total Factor Productivity (TFP) has Risen the Most in Developing Countries, with Brazil and China Leading the Way.

Region	Average TFP growth rate (% per year)				
	1961-70	1971-80	1981-90	1991-00	2001-14
All Developing Countries	0.69	0.93	1.12	2.22	2.21
Sub-Saharan Africa	0.17	-0.05	0.76	0.99	0.51
Latin America & Caribbean	0.84	1.21	0.99	2.3	2.74
Brazil	0.25	0.6	3.02	2.62	4.03
Asia (except West Asia)	0.91	1.17	1.42	2.73	2.78
China	0.94	0.67	1.71	4.1	3.05
West Asia & North Africa	1.4	1.66	1.63	1.74	1.88
All Developed Countries	0.99	1.64	1.36	2.23	2.44
United States & Canada	1.25	1.67	1.31	2.18	2.24
West and Central Europe	0.58	1.44	1.43	1.25	1.98
Transition Countries (former USSR and E. Eur.)	0.57	-0.11	0.58	0.78	2.28
World	0.18	0.6	0.62	1.65	1.84

High Income Countries

Total resources employed in agriculture in high income countries have been falling since, about 1980; TFP growth offset the declining resource base to keep output from declining. TFP growth has remained robust overall, but has slowed in some countries like Australia and the UK. Labour productivity has been rising much faster than land productivity and average farm size has increased—agricultural labour has been falling more rapidly than land used in agriculture.

Developing Regions

TFP growth in developing regions doubled between 1960-1990 and 1991-2009, from less than 1 per cent to over 2 per cent per year. Input growth has been slowing each decade but is still positive, enough to keep output growing at over 3 per cent annually for each of the last three decades. Two large developing countries in particular, China and Brazil, have sustained exceptionally high TFP growth over the past two decades. Several other developing regions—including Southeast Asia, North Africa, and Latin America—also registered accelerated TFP growth in the 1990s or 2000s compared with previous decades. The major exception is Sub-Saharan Africa, where long-run TFP growth has remained below 1 per cent per year.

Transition Countries

The dissolution of the USSR in 1991 imparted a major shock to agriculture in countries of the former Soviet Bloc. As these countries began the transition from centrally-planned to market-oriented economies, agricultural resources sharply contracted and output fell. Since, about 2001, output has begun to expand again, and it appears to be led by improvements in productivity. TFP growth, which was practically non-existent during the Soviet era, has taken off since, 2001. However, gross agricultural output in 2009 was still below Soviet-era levels in every region except Central Asia and Caucasia.

New research has measured agricultural TFP growth not only for all of the world's nations, but for the various states and provinces of large countries, namely, for China, the United States, Brazil, Australia, and Indonesia. This work shows that productivity is highly variable not only across countries, but within countries as well. In China, TFP growth has been very strong in coastal areas but slackens in the interior. Coastal states of Brazil have also experienced robust agricultural productivity growth. But unlike China, high TFP growth is also evident in some parts of the interior—like Mato Grosso in the Cerrado, now the main soybean and cotton producing state in the country. In Indonesia, productivity growth has been concentrated in recent years in the western and northern regions of the country—Sumatra and Kalimantan especially—where export commodities like oil palm and cocoa have been booming. In contrast, TFP growth has been low or stagnant in Java and the eastern provinces. This is a sharp departure from the "Green Revolution" decades of the 1970s and 1980s which disproportionately benefitted irrigated rice production, which is especially important in Java. In the United States, productivity growth has been moderately strong in the agriculturally important Corn Belt and Lake States, but low in Plains States, Appalachia and major horticultural producers like California and Florida. While Australian "broadacre" (dryland) agricultural TFP has stagnated nationally, this has primarily affected eastern and southern portions of the country.

The improved productivity growth performance in some sub-Saharan African countries. While a few countries appear to have raised their long-term agricultural TFP growth to over 1 per cent per year, others—like Angola—are simply recovering from earlier decades when they were at war. Sub-Saharan Africa remains perhaps the biggest challenge in achieving sustained, long-term productivity growth in its agricultural sector. It is also the region of the world with the highest poverty rate and in the coming decades it is facing the world's highest population growth rate.

DRIVERS OF AGRICULTURAL PRODUCTIVITY GROWTH

Perhaps the single most important factor separating countries that have successfully sustained long-term productivity growth in agriculture from those that have not is their national capacity in agricultural research and development (R&D). Countries that have built national research systems capable of producing a steady stream of new technologies suitable for local farming systems are generally the ones that have achieved the higher growth rates in agricultural TFP. Evidence finds that international and inter-state spillovers of agricultural knowledge are important sources of productivity growth, and that an important role of local R&D is facilitating the "capture" of these spillovers. Local R&D is often critical for adapting technologies developed elsewhere into useable technologies for local farming systems. Being actively engaged with foreign or international research institutions significantly raises returns to national agricultural research. While public-sector investments in agricultural R&D exhibited a slowing rate of growth in the 1980s and 1990s, the most recent evidence suggests that at least in developing countries this trend has reversed. In high income countries, some of the stagnation in public agricultural R&D spending has been offset by greater R&D investment by the private sector, although the willingness of the private sector to invest in agricultural R&D may in turn be dependent on continued advances in publicly-funded fundamental sciences.

In addition to R&D, new econometric evidence has identified a number of other factors that have contributed to cross-country differences in agricultural TFP. This can broadly be characterized as the "enabling environment" for the dissemination of new technologies and practices. These factors include policies that improve economic incentives for producers, stronger rural education and agricultural extension services, and rural infrastructure that improves access to markets. At the same time, economically disruptive "shocks," such as armed conflict and human or animal diseases—HIV/AIDS in Africa and avian flu in Asia—have seriously depressed agricultural productivity growth in some countries. Having a more favourable enabling environment compliments but does not substitute for research. Improving on these enabling factors raises the return to investments in agricultural R&D.

Future challenges to world food security, as in the past, do not appear to be related to technical constraints to raising agricultural productivity at the global level, but rather to uneven access to resources, technologies and food. Regions that have lagged behind the agricultural technology frontier, like much of Sub-Saharan Africa, have remained mired in poverty and food insecurity. These countries could follow the examples of agricultural success stories like Brazil and China, which invested heavily in agricultural research, made critical reforms to policies and institutions, and tapped into international sources of agricultural technology to raise their farmers' productivity, lower food prices for consumers, and stimulate economic growth. When a country's population shares broadly in these developments, it can have a major impact on poverty reduction and improving societal well-being.

AGRICULTURE GROWTH RATE

Agriculture Growth Rate in India GDP had been growing earlier but in the last few years it is constantly declining. Still, the Growth Rate of Agriculture in India GDP in the share of the country's GDP remains the biggest economic sector in the country. In the 1980s, Indian policymakers shifted their focus from food self sufficiency to generating additional income in rural areas as a means of tackling the problem of poverty, which was concentrated in rural areas. Acceleration of agricultural growth, with a special focus on improving the position of small farmers and extending the productivity revolution to non-irrigated areas was seen as a critical part of the strategy for poverty alleviation. This effort was supplemented targeted anti poverty programmes to address the needs of vulnerable groups who may not benefit sufficiently from general agricultural growth. India achieved considerable success with this approach in the 1980s. Growth of agricultural gross domestic product (GDP) accelerated to about 4.7 percent in the 1980s, compared with only 1.4 percent in the 1970s. This, agricultural growth, together with the beginning of economic reforms in the nonagricultural sector, pushed up the growth rate of overall GDP to around 5.8 percent in the period 1980-81 to 1989-90 compared with about 3 percent in the 1970s.

India's growth was disrupted at the start of the 1990s by a major balance of payments crisis which led to the adoption of an extensive process of structural reforms. It took time to regain momentum and it was only in 1993-94 that the economy got back on track, clocking an average growth rate of 6.8 percent in the three years 1993-94 to 1995-96. This acceleration in growth in the post reform period led policymakers to set a more ambitious GDP growth target of 8 percent a year for the Ninth Plan period (1997-98 to 2001-2002), to be supported by a growth rate of 4 percent a year in agriculture. The projected growth of 4 percent per year in agriculture was clearly inline with the average growth of 3.8 percent achieved in the period 1990-91 to 1996-97.

However, actual performance since the mid 1990s has been disappointing. Agricultural growth slowed to 2 percent a year in the Ninth Plan period, and overall economic growth was only 5.5 percent, well below the 8 percent target. Since agriculture accounted for about 25 percent of GDP, the shortfall of more than 2 percentage points in agricultural GDP growth compared with the target accounts directly for a shortfall of about half a percentage point in GDP growth. If the indirect effects of more rapid agricultural growth on other sectors are taken into account, the total impact on GDP growth may have been as much as one percentage point.

These shortfalls were known, when the Tenth Plan (covering the period 2002-03 to 2006-07) was formulated, but it was assumed that the poor performance of agriculture was due to temporary factors such as poor monsoons and depressed agricultural commodity prices in world markets following the East Asian meltdown. The Tenth Plan therefore adopted the same targets of 8 percent growth in GDP and 4 percent growth in agriculture. Experience in the first three years of the Tenth Plan period has sounded some alarm bells. GDP growth has averaged about 6.5 percent, but agricultural GDP in these years

(2002-03 to 2004-05) has grown by only 1.1 percent per year. The loss of dynamism in agriculture explains most of the shortfall in aggregate GDP growth.

Slower growth in agriculture also has direct implications for poverty reduction in rural areas. Official figures suggest that the incidence of poverty fell from 36 percent in 1993-94 to 26 percent in 1999-2000. The comparability of these numbers has been questioned because of recent changes (ostensibly improvements) in the methods for measuring consumption in household surveys, but there is a broad consensus that if corrections are made to ensure comparability, the percentage of the population in poverty has declined significantly, though less than in the official figures. However, even the official figures show less decline than what had been targeted, and this is undoubtedly a reflection of the slowdown in agricultural growth. Slow growth in agriculture is also at the root of growing evidence of distress in the farming community. Surveys show that a large percentage of farmers want to leave farming because they find it is no longer sufficiently profitable. The uncertainty associated with farming has also increased and farmers lack effective means of insuring against such risks. There are larger market uncertainties associated with new crops and poultry because of greater vulnerability because of falling ground water levels. There is evidence of increased indebtedness arising from the inability to cope with risks.

Recognizing these problems, the Government has undertaken a comprehensive review of the strategy for agriculture in order to come up with a new deal for agriculture and the rural economy in general. Remedial action will be needed on several fronts including increased public investment in irrigation and rural roads, better management of existing irrigation systems and of water resources in dry land areas, a strengthened agricultural research system and more effective extension, improvements in the production and distribution of certified seeds, improvements in the credit delivery system, and innovative steps in marketing and contract farming to support the diversification of Indian agriculture.

IRRIGATION

Water is a critical constraint on raising agricultural productivity and much of the success of the Green Revolution came from improved productivity in areas of assured irrigation provided through canals or (much more significant) through ground water utilization. The scope for expanding irrigation through large and medium scale projects has yet to be fully exploited. Out of the total of 59 million hectares that could be irrigated through such projects only 40 million hectares have been irrigated. The slow pace of exploitation of irrigation potential is due to lack of resources in state governments and the tendency to spread available resources thinly over too many projects. Additional public investments in this area are therefore essential for early utilization of the potential. Effective maintenance of the existing system of canal irrigation also suffers because the irrigation departments of the states lack resources. This in turn is because water charges are kept too low, covering only 20-25% of the operations and maintenance cost of the system in most states. Poor maintenance leads to loss of water through seepage, with the result that water use efficiency is very low – around 25 to 40 percent instead of 65 percent that should be attainable. Low water charges also encourage highly water intensive crops at the upper end of the canal network, leaving tail-end portions starved of water.

The solution lies in rationalization of water rates to ensure adequate financial resources to cover maintenance and resort to participatory irrigation management to give farmers a stake in the operation and maintenance of the system. Some interesting experiments in these have promise. Maharasthra has recently established a Water Regulatory Authority to set water charges in a non-political manner. Several states are also experimenting with involving water user associations (WUAs) in the operation of the canal systems. Ideally the WUAs should be empowered to collect water charges and to retain a portion of the collection to maintain the portion of the distribution network operating in their area.

Ground water utilization played a major role in expanding irrigation in the 1980s but uncontrolled exploitation of groundwater has led to serious depletion of the water table in many parts of the country. Overexploitation is encouraged by the policy of massive under pricing of electricity for agricultural use, with a few states having made electricity for farmers completely free. Even where it is not free, the charge for electricity is a fraction of the average cost, and is not based on metered use. Instead there is a fixed charge for presumed usage, based on the capacity of the pump, an arrangement which implies that the marginal cost of electricity for pumping ground water is zero. Under pricing canal water and electricity are clearly highly distortionary, given the need to conserve water use. They are also distributionally unfair because the benefits of under priced water accrue disproportionately to upper end farmers whereas under priced power enables those able to afford larger pumps to lower the water table denying water to farmers who can only afford shallow wells.

The investment requirements of irrigation are massive. Completion of all unfinished projects alone is estimated to cost approximately US$ 20 billion. In addition, provision must be made for new irrigation projects (large, medium and small), which together will require about US$ 45 billion. The total requirement is therefore about US$ 65 billion.

TYPES OF CREDITS

The Credit requirements of agriculture are of three types viz.:

- Short-Term

- Medium-Term
- Long-Term (LT)

Long-Term Credit

The period of long-term credit is generally 5 to 20 years or even more in some special cases. In any industry, long-term investment is necessary, to create permanent assets which give returns over a period of time. The permanent investment is not only necessary for a particular industry but even for the country. Because for continuity of production and progress of the country. This applies to agriculture also. In Agriculture, long-term investment comprises of sinking well, land leveling, fencing and permanent improvements on land purchase of big machinery like tractor with its attachments including trolleys, establishment of fruit orchard of mango, cashew, coconut, sapota (chiku), orange, pomegranate, fig, guava, etc.

There are many other items of long-term capital investment. Investment once made in the beginning continuous to give returns over a long period. Fruit orchards particularly do not give any income in the first 4-5 years as incase of other seasonal crops. So the expenditure incurred in the first 4-5 years becomes a capital cost.

All the long-term investments mentioned above require large amounts of funds. Although they have good potential to give returns in future, individual farmers have no financial capacity to make such costly investments from their own funds because they have no savings or very little savings. Therefore, they have to resort to bank borrowing to meet their such needs. The financial criteria terms and conditions procedures of granting L.T. loans are altogether different from short-term loans: Even the bank or agency providing LT loans is separate due to its particular mode or system of raising capital and gain.

Land Development Banks

The special banks providing LT Loans are called Land Development Banks (LDB). The history of LDB's is quite old. The first LDB was started at Jhang in Punjab in 1920. But the real impetus to these banks was received after passing the Land Mortgage Banks Act in 1930's (LDB's were originally called Land Mortgage Banks). After passing this Act LDB's were started in different states of India.

Structure

These Banks have two-tier structure:

- Primary Land Development Bank at district level with branches at taluka level.
- Control or State Land Development Bank. All primary Land Development Banks are federated into Central Land Development

Bank at the State Level. In some States, there is "Unitary structure" wherein, there is only one State Land Development Bank at the state level operating through its branches and sub-branches at district and below levels.

Raising Funds

The main function of raising funds is carried out be the Central or State Land Development Bank which can really deal with the money market of the country effectively and advance loans to primary LDB's.

The sources of funds of State LDB's are:

- Share capital.
- Issue of debentures
- Loans from NABARD
- Reimbursements of subsidies from the Government.
- Other funds.

Issue of debentures is the main source of funds for the LDB's. Debentures is a 'Bond' conveying and acknowledging the debt and also containing the provision of promise for payment of interest at stipulated rate and return of the principal amount. The period of debentures varies from 7 to 15 years. As LDB's require funds of longer duration to advance LT loans to borrowers, the debenture is a convenient instrument of raising funds. Because it guarantees that funds will remain with the Banks for a specified period.

There are three types of debentures:

- Regular debentures
- Rural debentures
- Special development debentures.

These debentures are mostly purchased by financial institutions like LIC, Commercial Banks, Co-op. Banks, NABARD, and State Governments. As there is limited response from the public. The State Government give incentive subsidies for many development activities by individual farmer including purchase of tractor. The amounts of subsidies are reimbursed to the LDB's.

Interest Rate

The rates of interest for LT Loans are generally low and within the paying capacity of farmers. They are around 11 to 12%.

Loan Procedure

The Branch offices receive applications from the prospective borrower. Then Agricultural Finance Officer or Inspector scrutinises these applications, they visit places of the application and ascertain the purpose of borrowing, verify the genuineness of the proposal and it economic viability, repaying ability of the farmers, adequacy of security, etc. After completing those formalities, the loan is granted by the appropriate authority at appropriate level depending upon the delegation of powers by the Banks.

CROP LOAN

Crop loan is a short term credit and is generally obtained from primary credit co-op. Society of a village or also from commercial bank. The period of loan is about one year except for sugarcane for which the period is 18 months.

There are two criteria for granting crop loan:

- One third of gross value
- Cost of cultivation.
- One third of gross value approach takes into account the yield and price of the crop, its cost of cultivation and family expenditure. If the gross value is more, more amount of loan becomes available. For example Rice.

	I	II
Yield (Q.)	20	25
Price (₹/Q)	400	400
Gross value (₹.)	8000	10,000
One third (₹.)	2700	3330

- Thus in second situation farmer is entitled for ₹ 3330 per hectare which is higher than in the first situation. Thus this method takes into account the productive aspect of a crop.
- In cost of cultivation, direct paid-out costs are only considered. They include items, like seeds, manures, fertilizers, pesticides, diesel/electricity, hired labour etc. In this approach, it is expected that all direct costs to be incurred by the farmer should be covered and accordingly he should get adequate credit. If the cost of all these items of input is ₹ 3500/-. If the loan is granted according to first approach, then the amount which is short, is spent by the farmer from his own funds. Since crop loan is for one season, its recovery is made in one installment after the harvest of the crop. Crop loan is an annual requirement and farmer has to borrow fresh loan for new crop season every time. Therefore, he has to repay the earlier loan with interest within stipulated time. Since this loan is required every season/every year, the procedure of getting this loan is simple and convenient and it is made available by the District Central Co-op. Banks through the village Co-op. Credit Society. So the farmer gets his loan in the village itself. If the loan is to be taken from commercial bank, it is available from the nearby branch of the commercial bank. As for security, the farmer has to offer his land as a security. There is a three tier structure providing crop-loans through co-operative institutions.

– Appex Bank- State Co-op. Bank.
– District Central co-op. Bank
– Village co-op. Credit Society.

Crop-loan is the most important need of the farmer to increase and maintain his productive ability. With the help of this loan amount, he can purchase modern costly inputs and adopt new technologies on his farms. So through these loans co-operative banks play important role in the development and prosperity of agriculture. Among the various types of bank loans to agriculture, the share of crop loan is the highest.

Agricultural Loans

Agricultural loans are available for a multitude of farming purposes. Farmers may apply for loans to buy inputs for the cultivation of food grain crops as well as for horticulture, aquaculture, animal husbandry, floriculture and sericulture businesses. There are also special loans to finance the purchase of agricultural machinery such as tractors, harvesters and trucks. Construction of biogas plants and irrigation systems as well as the purchase of agricultural land may also be financed through special types of agricultural finance.

COOPERATIVE AGRICULTURAL BANK

- National Bank for Agriculture and Rural Development or NABARD- is responsible for refinance disbursement to commercial banks, State cooperative banks, State cooperatives, rural development banks, Regional Rural Banks (RRBs) and other eligible financial institutions. It also sanctions money through its Rural Infrastructure Development Fund for projects covering irrigation, rural roads and bridges, health and education, soil conservation and drinking water schemes. NABARD also offers a Kisan Credit Card Scheme and crop loans under the Rashtriya Krishi Bima Yojana.

  Banks and RRB's introduced the Kisan Credit Card Scheme of NABARD in their areas of operation. In this scheme eligible farmers are provided with a Kisan Credit Card and a passbook or card-cum-pass book. The revolving cash credit facility allows any number of withdrawals and repayments within the limit. This limit is fixed on the basis of operational land holding, cropping pattern and the scale of finance. Sub-limits may be fixed at the discretion of banks. This Kisan Credit Card is valid for 3 years subject to annual review. As incentive for good performance, credit limits may be enhanced to take care of increase in costs, change in cropping pattern, etc. Each drawl should be repaid within a maximum period of 12 months. Conversion or rescheduling of loans is allowed in case of damage to crops due to natural calamities. Security, margin, rate of interest and other details are fixed according to RBI norms.

- Bihar State Co-operative Bank Limited (BSCB)- Offers a range of loans and financial schemes to agriculturalists.
- Haryana State Co-operative Apex Bank Limited (HARCOBANK)- The bank offers crop loans, Kisan Credit Cards, cash credit against hypothecation of stocks and interim finance by way of cash credit.
- National Federation of State Co-operative Banks Limited (NAFSCOB)- This federation offers a range of agricultural loans through member State Cooperative Banks, District Central Cooperative Banks and Primary Agricultural Cooperative Societies.
- Orissa State Co-operative Bank Limited (OSCB)- The bank has introduced Kisan Credit Cards in the S.T. Cooperative Credit Sector. It also organizes seminars on agri finance. OSCB has 17 Central Cooperative Banks and around 810 mini banks in different districts of Orissa.
- Repatriates Co-operative Finance and Development Bank Limited- This bank does not have any specific agricultural loan, but offers a range of financial products that can be accessed by people who wish to develop agriculture and related activities.
- Punjab State Cooperative Agriculture Development Bank Ltd- Initially, the bank only gave farmers loans to pay off old debts and purchase land. Today, the bank provides loans for various purposes like improvement of alkaline and saline lands, purchase of tractors, installing tube wells and other modern agricultural equipment. It also offers financial schemes for poultry development, dairy development, horticulture, floriculture, sheep rearing and inland fisheries.
- Andhra Pradesh State Cooperative Bank Limited (APCOB)- has a loan portfolio that covers crop loans, medium term loans and long term loans for agricultural purposes. It also supports government sponsored District Rural Development Agency projects through IRDP loans and cooperative sugar factories, spinning mills, weaver's societies, employees' cooperative credit societies and other organizations. APCOB has also extended finance to apex cooperative institutions in the State such as APCO, MARKFED and GCC.

NATIONALISED BANKS

- *Allahabad Bank*: Offers the Kisan Credit Card and Kisan Shakti Yojana Scheme. The Kisan Credit Card is a unique scheme for farmers through which they can draw a cash loan for crop production as well as domestic needs from the card-issuing branch within the sanctioned limit. The Kisan Shakti Yojana provides farm investment credit, as well as personal/domestic loans including repayment of debt to moneylenders. The permissible loan limit will be 50 per cent of the value of land or 5 times the net farm income, whichever is lower, less the outstanding amount, if any, in Agril.

- *Andhra Bank*: Provides facilities to farmers like AB Kisan Vikas Card, AB Pattabhi Agricard, AB Kisan Chakra, rural godowns, agri clinics, agri service centres, self help groups and solar cookers. They also provide other schemes such as Kisan Sampathi, tractor financing, Kisan Green Card, Surya Sakhti and loans to dairy agents.
- *Bank of Baroda*: Offers farmers the Baroda Kisan Credit Card. It also has schemes for the purchase of agricultural implements, heavy agricultural machinery like tractors, irrigation and other infrastructure. Bank of Baroda also finances the development of agri industries like horticulture, sericulture, fisheries, dairy and poultry.
- *Bank of India*: Has a Kisan Credit Card Scheme that helps farmers raise short-term funds for agriculture and other farm-based activities, on an on-going basis, with very flexible and friendly repayment terms. It also offers an agricultural loan for development of agriculture related industries, purchase of machinery and other agricultural purposes.
- *Bank of Maharashtra*: Offers agriculturists a Maha bank Kisan Credit Card and financial schemes for digging new wells, purchasing harvesters, livestock, vehicles and land. Repayment terms for different agricultural loans range from three to fifteen years.
- *Canara Bank*: Provides Kisan Credit Cards. Limits up to 50,000 have no margin while those above 50,000 have a margin of 15 to 20 per cent. Other than this, Canara Bank provides a wide array of financial schemes for different agricultural purposes.
- *Central Bank of India*: The Central Kisan Credit Card is a credit service provided to farmers on the basis of their holdings for purchasing agricultural inputs. Only those farmers having a good track record for the past 2 years with the bank as a borrower or depositor and who are not defaulters to any credit institution would be considered for loans.
- *Corporation Bank*: Offers a range of loan schemes to farmers. They are the Corp Gram Mitra Yojana, Corp Arthias Loan Yojana, Corp Kisan Tie-Up Loan Scheme, Corp Kisan Farm Mechanisation Scheme and Corp Kisan Vehicle Loan Yojna.
- *Dena Bank*: Dena Bank has sponsored 2 Regional Rural Banks namely Dena Gujarat Gramin Bank in Gujarat and Durg Rajnandgaon Gramin Bank (DRGB) in Chhattisgarh. The bank has set up a Rural Development Foundation for training unemployed youth in rural areas. Other financial schemes of the bank are the Dena Swacch Gram Yojana, Dena Kisan Gold Credit Card Scheme and the Dena Bhumiheen Kisan Credit Card Scheme.
- *Indian Bank*: Has a wide range of schemes for agriculturalists such as Swarojgar Credit Card, Gramin Mahila Sowbhagya Scheme, Kisan

Bike Loan Scheme, Yuva KisanVidya Nidhi Yojana and Indian Bank Kisan Card Scheme.

- *Indian Overseas Bank*: Offers agri business consultancy services that include conducting feasibility and market studies, preparation of detailed project reports and formulation of rehabilitation packages for sick agro units.
- *Oriental Bank of Commerce*: It has two agricultural projects - the Grameen Project and the Comprehensive Village Development Programme. The Grameen Project involves disbursing small loans ranging from ₹ 75 onwards to mostly women. Training is also provided in villages in using locally available raw material to produce pickles and jams. The Comprehensive Village Development Programme focuses on providing an integrated package of rural finance to villagers to build up their village.
- *Punjab and Sind Bank*: Offers a range of financial schemes for farmers like the Zimidara Credit Card, tractor finance scheme, drip irrigation scheme, Kheti Udyog Khazana Yojana, vermi composting scheme, horticulture clinic and private veterinary clinic with dairy unit scheme.
- *Punjab National Bank*: This bank has a special website called PNB Krishi for agriculturalists. It gives details on crop practices, plant protection, farm machinery, market prices and other farming news and activities. The website also provides a list of financial schemes offered by Punjab National Bank on production credit, investment credit, composite loans, animal husbandry and farm mechanization.
- *Syndicate Bank*: Offers a wide range of agricultural loan products such as the Synd Jai Kisan Loan Scheme, Jewel Loan Scheme for Agriculture, Syndicate Farm House Scheme, Finance for Hi-tech Agriculture, Development of Irrigation Infrastructure scheme, Syndicate 2/3/4 Wheelers Scheme and the Syndicate Kisan Credit Card (S.K.C.C).
- *UCO Bank*: This Bank provides the UCO Hirak Jayanti Krishi Yojana to meet the long-term credit needs of the farming community in rural areas for agriculture, allied activities as well as for personal purposes. Only farmers below 60 years are eligible to apply. Minimum quantum of the loan is ₹25,000/- and the maximum is ₹5 lakhs.
- *Union Bank of India*: Facilities provided to farmers include Kisan ATM Cards and special Kisan ATM Machines. These ATM's are easy to operate and do not require farmers to have a high level of literacy. They are voice enabled in the local language, have a touch screen monitor and work on a bio-metric authentication system like finger print verification.

- *United Bank of India*: The range of financial schemes offered to agriculturalists include the United Krishi Laghu Paribahan Yojana, United Krishi Sahayak Yojana, United Gramya shree Yojana, Gramin Bhandaran Yojana and the United Bhumiheen Kisan Credit Card.
- *Vijaya Bank*: This bank offers one comprehensive financial scheme known as the Vijaya Krishi Vikas (VKV) Scheme. This scheme provides a simple package to farmers to meet entire agricultural credit requirements such as crop production, investment credit and consumption credit. All farmers including owners, tenant cultivators, leased land farmers and sharecroppers are eligible for this scheme.

SCHEMES FOR AGRICULTURE FINANCE

SBT Kisan Gold Card Scheme (General purpose Agriculture Term Loan)
Eligibility:

- Farmers having good track record of repayment for the last two years.
- Farmers who have closed their loan account without default and not our current borrowers.
- Farmers who have defaulted in repayment but closed the Loan within the stipulated repayment period.
- Farmers who are maintaining deposits with the Bank.
- Good borrowers of other banks provided they liquidate their dues with other banks.
- Good farmers who have not availed loans from any bank.
- *Purpose*: The borrower is at liberty to utilize 50% of the amount for any purpose, including consumption purpose and purchase of land.
- *Amount of Loan*: The amount of loan is limited to five times the annual farm income including income from allied activities or 50% of the value of the land offered as collateral security, whichever is less, subject to a maximum of ₹10 lakh.
- *Rate of Interest*: Interest rate ranges from 1% below PLR.
- *Security*: Hypothecation of crops and assets, if any, created out of bank finance and existing movable assets such as milch animals, pump sets etc. The loan will be secured by equitable mortgage of properties worth double the loan amount, or term deposit receipts, LIC policies of adequate surrender value, NSCs completed lock in period or more etc.
- *Disbursement*: Cash disbursals are allowed to the full extent of the credit limit.
- *Repayment*: The repayment period shall be 10 years. The due date of the installment shall be fixed in such away to coincide with the date of generation of income.

KISAN CREDIT CARD SCHEME

- *Eligibility*: All agriculturists who are in need of short term production requirements. ATM facility and Personal Accident Insurance Scheme for life up to ₹ 50000 and permanent disability cover up to ₹ 25000 is available on request.
- *Purpose*: To provide hassle free short-term credit to farmers on the basis of their land holdings for purchase of inputs and draw cash to meet their production needs. *i.e.* Cultivation expenses including allied activities with a consumption component.
- *Amount of Loan*: To be fixed on the basis of operational holdings and scale of finance with consumption component 15% (maximum ₹ 10000/-) of production credit. The scale of finance to farmers who own cultivated land below one acre will be at the rate of ₹40000/- (on pro rata basis) and farmers who own more than one acre with intensive farming of land be given at the rate of ₹37500/- per acre and part thereof.
- *Rate of Interest*: Interest rate ranges from 2.50% below to 1.50% above BPLR for various limits.
- *Repayment*: Running Cash Credit account for 36 months subject to annual review and total annual credit should exceed annual debit.

HOMESTEAD FARMING

- *Purpose*: A scheme for financing farmers practicing mixed cropping/ intercropping along with allied activities to enable them to undertake cultivation of various crops in a more integrated way. The scheme provides the farmers with sufficient working capital required for their homestead farming (Mixed cropping along with allied activities) by fixing scale of finance based on land holding to meet the cost of entire farming activities.
- *Amount of Loan*: The farmers who own cultivated land below one acre be given the scale of finance on pro rata basis at the rate of ₹ 40000/- and farmers who own more than one acre of land be given at the rate of ₹ 37500/- per acre and part thereof.
- *Rate of Interest*: Interest rate ranges from 2.50% below to 1.50% above BPLR for various limits.
- *Repayment*: The facility will be sanctioned as an Agriculture Cash Credit limit (In case of Kisan Credit Card running cash credit).

LOAN FOR ESTATE PURCHASE

- *Eligibility*: The estate should be either in yielding stage with the crops in its prime yield age or capable of being developed in to a viable unit. The yield/ net income of the estate should be sufficient to liquidate the proposed loan and interest accrued with in a period of 7 to 10 years. The proposed estate should be free from encumbrance and entire property should be offered as security to the loan.

- *Purpose*: To encourage those who prefer to settle down in agriculture and are in the look out of good/viable estates for purchase and also to improve production in agriculture.
- *Amount of Loan*: The quantum of loan that will be considered for sanction will be 75% of the registered value or 50% of the market value whichever is low. In exceptional cases 80% of the registered value or 50% of the market share whichever is low is also considered. The loan for the development of the estate like land development including working capital can also be sanctioned.
- *Rate of Interest*: Interest rate same as BPLR.
- *Repayment*: Repayment of loan will be in quarterly/half yearly/yearly installments depending on the harvest of the crops and the loan shall be repaid within a maximum period of 7 to 10 years.

SCHEME FOR FINANCING FARMERS FOR PURCHASE OF LAND FOR AGRICULTURAL PURPOSES

- *Eligibility:* Small and Marginal farmers - land maximum up to 5 acres of non-irrigated land or 2.5 acres of irrigated land including the land purchased under the scheme. Tenant, sharecropper and landless agricultural labourers with a good record of prompt repayment of our loans for the last 2 years are also eligible.
- *Purpose*: To finance small and marginal farmers, share croppers, tenant cultivators for purchasing land to expand activities and to make existing small and marginal units economically viable to bring fallow lands and waste lands under cultivation to step up agricultural production as well as productivity also to finance share croppers/ tenant farmers to enable them to diversify farming activities to allied areas to increase their income.
- *Amount of Loan*: Maximum loan under the scheme towards land cost shall not exceed ₹ 5 lakh. Cost of development/economic activity shall be financed under the bank's other financing schemes.
- *Rate of Interest*: Interest rate ranges from 1.75% below to 2.00% above BPLR for various limits.
- *Repayment*: Repayment of the loan will be 7 to 12 years in half yearly/ yearly installments with maximum of 24 months moratorium period. Gestation period/repayment due dates etc. will be fixed according to income generation from the activity.

SCHEME FOR CULTIVATION OF MEDICINAL PLANTS

- *Eligibility*: All agriculturists are eligible.

- *Purpose*: Scheme for financing cultivation of 22 medicinal plants cultivated extensively and also in great demand in the local as well as foreign market.
- *Amount of Loan*: Depending on the area of cultivation/ project cost
- *Rate of Interest*: Interest rate ranges from 1.75% below to 2.00% above BPLR for various limits.
- *Repayment*: Repayment should coincide with harvesting and marketing or at the time generation of income from the scheme.

SCHEME FOR CULTIVATION OF VANILLA

- *Eligibility*: All agriculturists are eligible.
- *Purpose*: Scheme for financing cultivation of Vanilla, a cash crop, gaining ground in the State of Kerala.
- *Amount of Loan*: Amount of finance will be ₹250000 per hectare for pure crops and ₹210000/- per hectare for intercrop.
- *Rate of Interest*: Normal rate of interest as applicable to ATL.
- *Repayment*: The loan shall be repaid within a period of 7 years, in yearly installments. Farmer's eligible for two years gestation period and interest is repayable on the 3rd and 4th year and the principal from the 5th to 7th year.

SBT RAINWATER HARVESTING SCHEME

- *Eligibility*: Farmers having land holding of 0.50 acre or more are eligible to be considered for finance under this scheme.
- *Purpose*: Scheme envisages construction of low cost tanks for collecting and storing rainwater and using it for irrigation, by siphon arrangement, utilizing gravitation flow or by installing motor pump.
- *Amount of Loan*: Maximum amount of finance will be ₹88000/- per acre. Scheme can be adopted in smaller areas also by reducing the cost proportionately.
- *Rate of Interest:* Interest rate ranges from 1.75% below to 2.00% above BPLR for various limits.
- *Repayment*: Repayment based on the income generated from the crops raised and cropping pattern. The maximum period eligible for repayment is 8 years in annual instalments.

PRODUCE MARKETING LOAN (ADVANCE AGAINST WAREHOUSE RECEIPT)

- *Eligibility*:
  - Farmers/ traders depositing farm produce in the warehouses of the central/ state warehousing corporations.
  - Scheme will be operative in Karnataka, Andhra Pradesh, Tamilnadu and Kerala.

- *Purpose*:
  - To protect the farmers from the compulsion to sell their produce immediately after harvest of produce despite an adverse market.
  - To finance farmers and traders against warehouse receipt.
- *Amount of Loan*: 70% of the value of the warehouse receipt, valued at the market value or 70% of the market price advised by Agri. Dept, HO whichever is less.
- *Rate of Interest*

  Farmers:
  - Up to `3 lakh - 3.50% below PLR 9.50%
  - Above `3 lakh - 2.50% below PLR 10.50%

  Traders:
  - 2.50% below PLR 10.50% (Irrespective of the limit)
- *Repayment:* On demand/6 months which can be extended up to 12 months subject to satisfactory shelf life/market condition.

AGRI-LOAN TO NON-RESIDENT INDIANS

- *Eligibility*: Agricultural advances are available to the resident family members (means spouse, father, mother, brother, sister etc.) of Non-Resident Indians for land-based activities in respect of the land held by them in India subject to:
  - The loan should be need based and the total land holding of the Non-Resident Indian, in individual name or jointly with others, should not exceed 5 ha.
  - The loan amount shall not be used for acquiring any additional land.
- *Purpose:* To finance farmers only for land-based activities and to carryon agricultural activities on the existing land.
- *Amount of Loan*: The maximum amount of the loan will be need based.
- *Rate of Interest*: Interest rate ranges from 2.50% below to 1.50% above BPLR for various short-term limits and from 1.75% below to 2.00% above BPLR for various long-term limits.
- *Repayment*: The loan can be repaid out of the income generated from the agricultural activities or remittances from abroad or by debit to their NRE/NRO/FCNR accounts.

MINOR IRRIGATION

Projects with cumulative command area of less than 2000 ha are called minor irrigation projects:

- *Eligibility*: The beneficiary should have a minimum of 50 cents of land to be brought under irrigation to ensure viability and repayment of loan.
- *Purpose*: Scheme for developing irrigation potential, Minor Irrigation, Installation of Pump set Drip Irrigation etc.

- *Amount of Loan*: As per the project submitted.
- *Rate of Interest*: Interest rate ranges from 1.75% below to 2.00% above BPLR for various limits.
- *Repayment*: The loan shall be repaid within a period of 9 years, in yearly installments.

FARM MECHANISATION

Loan for Farm Mechanisation, Purchase of tractors, Power Tillers, etc.

- *Eligibility*:
  - Tractors with engine capacity up to 35 HP – The applicant should own/ cultivate six acres of perennially irrigated land.
  - Tractors with engine capacity above 35 HP – The applicant should own/ cultivate eight acres of perennially irrigated land.
  - Power Tillers – the applicant should own/ cultivate four acres of perennially irrigated land.
- *Purpose*: To purchase tractor/power tillers for agricultural activities.
- *Amount of Loan*: Amount of advance will be the investment cost of tractor/ power tiller and implements less margin @15%.
- *Rate of Interest*: Interest rate ranges from 1.75% below to 2.00% above BPLR for various limits.
- *Repayment*: The period of repayment shall be 9 years for tractors and 7 years for power tillers.

AGRICULTURE GOLD LOAN

- *Eligibility*: All individual farmers undertaking cultivation or other activities including allied activities are eligible for short-term finance.
- *Purpose*: To meet genuine credit requirements of farming including allied activities, repairing of equipments and consumption needs etc.
- *Amount of Loan*: The eligible loan amount should be assessed based on the area under cultivation, crops(s) raised, scale of finance and not in relation to the value of gold offered as security.
- *Rate of Interest*: Interest rate ranges from 2.50% below to 1.50% above BPLR for various limits. For working capital loans like ACC/KCC/ AGL up to ₹ 3 lakh interest at the rate of 7% is extended as per RBI guidelines subject to the periods stipulated by RBI and beyond that normal rate will apply.
- *Repayment*: As applicable to Agri. Cash Credit accounts depending on the duration of crops raised and harvesting period and income generation, subject to a maximum period of 12 months. The account has to be closed at the end of the repayment period.

SCHEME FOR DEVELOPMENT/ STRENGTHENING OF AGRI. MARKETING INFRASTRUCTURE, GRADING AND STANDARDIZATION

- *Eligibility*: Scheme shall be available to individuals, groups of farmers/ growers/ consumers, partnership/partnership firms, NGO's, SHG, Companies, Corporations, Cooperatives, Co-marketing Federations, Local Bodies, etc.
- *Purpose*: For development of agricultural marketing operations including strengthening of infrastructure, techniques of preservation, storage, etc.
- *Amount of Loan*: As per the project.
- *Rate of Interest*: BPLR irrespective of credit size.
- *Repayment*: Adequate long-term repayment period according to the project.

CONSTRUCTION/ RENOVATION/ EXPANSION OF RURAL GODOWN

- *Eligibility*: The project for construction of rural godowns can be taken up by Individuals, Farmers, Group of farmers/growers, Partnership/ Proprietary firms, NGOs, SHGs, Companies, Corporations, Co-operatives, Federations, Agricultural Produce Marketing Committees, Marketing Boards and Agro Processing Corporations.
- *Purpose*: To create scientific storage capacity with allied facilities in rural areas to meet the requirements of farmers for storing farm produce, processed farm produce and agricultural inputs.
- *Amount of Loan*: As per the project.
- *Rate of Interest*: As applicable to advances under SIB/ CandI segments will be charged.
- *Repayment*: Adequate long-term repayment period, not less than 5 years including a grace period of one year.

Bibliography

A J Grove and G E Newell: *Animal Biology*, Shubhi Publication, Delhi, 2007.

Ashok Kumar Sharma: *Animal Biochemistry*, Random Publication, Delhi, 2012.

C. LLoyd Organ: *Animal Behaviour*, MJP Publishers, Delhi, 2012.

C.V. Singh: *Animal Breeding and Genetics*, New India Publishing Agency, Delhi, 2015.

Chandra Deo Narain Singh: *Advanced General Pathology of Animals*, IBDC Publication, Delhi, 2010.

D. Gopalakrishna Rao: *A Text Book on Systemic Pathology of Domestic Animals*, IBDC Publishers, Delhi, 2010.

D. Gopalakrishna Rao: *A Text Book on Tumors of Domestic Animals*, International Book, Delhi, 2004.

Deepa H. Dwivedi, S.K. Dwivedi and S. Gupta: *Agro-Animal Resources of Higher Himalayas*, Satish Serial Publication, Delhi, 2010.

Elle Matthews: *Animal Adventures in Africa : Serengeti Safari*, Unicorn Books, Delhi, 2010.

Florence Periera Raja: *Animal Biotechnology*, Wisdom Press, Delhi, 2013.

Ghulam Mohyuddin Wani: *Animal Agriculture Biotechnology: Biotechnologies, Policy, Planning, Reviews and Visions*, Satish Serial Publication, Delhi, 2009.

J A Bierens De Haan: *Animal Behaviour*, Reprint Publication, Delhi, 2004.

Johnson Stanley: *Animal and Bird Pest Management in Agricultural Land*, Daya Publishing House, Delhi, 2015.

Lata Bhattacharya: *Animal Biochemistry*, Discovery Publication, Delhi, 2010.

M M Ranga: *Animal Biotechnology*, Student Edition, Delhi, 2007.

M.M. Ranga: *Animal Behaviour*, Agrobios Publication, Delhi, 2002.

Mahendra Kumar: *Animal Byproducts Utilization Through Semi-Moist Rendering*, Daya Publication, Delhi, 2007.

P Ramadass: *Animal Biotechnology : Recent Concepts and Developments*, MJP Publication, Delhi, 2008.

P. Kanakaraj: *A Text Book of Animal Genetics*, International Book Distributing Company, Delhi, 2007.

Prasum Tyagi: *A Textbook of Animal Physiology*, Dominant Publication, Delhi, 2010.

R D Sharma: *An Introduction to Animal Behaviour*, ABD Publication, Delhi, 2004.

R S Chauhan and Kuldeep Dhama: *Aflatoxicosis in Animals and Its Public Health Significance*, International Book, Delhi, 2008.

R. K. Saxena and Sumitra Saxena: *Animal Adaptations: Evolution of Forms and Functions*, MV Learning an imprint of Viva Books, Delhi, 2015.

S.P. Tiwari, S. Rajagopal and Usha Rani Mehra: *Analytical Techniques in Animal Nutrition*, Satish Serial Publication, Delhi, 2012.

T.N. Ananthakrishnan and K.G. Sivaramakrishnan: *Animal Biodiversity : Patterns and Processes*, Scientific Publication, Delhi, 2006.

Udai Veer Singh: *Animal Biochemistry*, Sonali Publications, Delhi, 2011.

William J.A. Payne and R. Trevor Wilson: *An Introduction to Animal Husbandry in the Tropics*, Wiley, India, 2013.

Index

H

I

L

M

N

O

P

R

S

T